BUILDING
Barbecues
▣ PATIOS, PERGOLAS & DECKS ▣

BAY BOOKS PRACTICAL DESIGNS FOR LIVING

BUILDING
Barbecues
PATIOS, PERGOLAS & DECKS

Rob and Sue Whelan

BAY BOOKS
Sydney & London

Published by Bay Books
61-69 Anzac Parade
Kensington NSW 2033 Australia

Publisher: George Barber
Text copyright © Rob Whelan and
Sue Whelan

National Library of Australia
Card Number and ISBN 1 86256 115 X

The Publisher wishes to thank the
following for their assistance with
additional photography for this book:
Australian Picture Library — page 100.
Colonial Castings — page 96.
PGH Bricks — pages 18, 21, 44.
Sunproof Outdoor Furniture — page 77.
Sydney Brick Paving Company — page 47.
Weldon Trannies — pages 41, 60, 65, 76, 90.
Rob Whelan — pages 19, 46, 47, 50.

Photography: Ray Joyce
Designed by KJ Graphics
Typeset in Plantin by Savage Type,
Brisbane
Printed in Singapore by Toppan

CONTENTS

INTRODUCTION

A sunny day, a mouth-watering steak, a glass of wine and good company. It all adds up to that great institution — the barbecue. Blessed with a mild climate, long summers and an incredible choice of superb meat and fish, it's easy to understand why informal gatherings around a barbecue have become so popular.

Casual, outdoor entertaining has reached such acceptance, even in the evenings, that there are few homes that don't own some sort of barbecue either constructed in the garden or tucked away on wheels in the shed or garage.

What started out as a simple grill over a metal pan has now reached great heights of sophistication in design and performance. Thousands of portable units are sold each year while almost as many people decide to take up tools and build their own.

Building Barbecues is a comprehensive guide to designing and building a total barbecue entertaining area in your garden or outdoor living space. More than just a set of plans for constructing the cooking unit, this book will show you how to successfully complement your barbecue with paving or decking, a pergola, landscaping, lighting and furniture. You'll create a total outdoor recreational area that will be a pleasure to live with and a joy to entertain in.

Too often, barbecues are located in an obscure corner of the garden; a location rarely ideal for entertaining or relaxing. We'll show you the attractive alternatives. How to integrate the barbecue into the overall look of your house and garden. How to allow for the sun and wind. How to avoid the hassles associated with smoke, fumes, poor location and burnt sausages. And how to avoid the hazards of lighting a fire in the open. There are loads of great ideas plus attractive, colour illustrations to help you visualise it all happening in your own backyard. After you've decided on a basic plan, there are easy instructions to help you get started.

You'll see at a glance what's involved in designing, building and creating a highly efficient cooking unit within a pleasing recreational environment, whether you plan to do it yourself or hire professionals. *Building Barbecues* has been compiled for the do-it-yourselfer; someone who wants to do the whole job or at least play an active part in designing and overseeing the work from start to finish.

However you decide to work it, it's certain your new barbecue will become the focal point of your outdoor eating and entertaining, giving years of pride and pleasure to you, your family and friends.

1

FIRST DECISIONS

*Ten Quick Questions • Counting the Cost •
To Buy or Build? • Which Fuel?*

Before you order a truckload of expensive building materials it's a good idea to have a long, hard think about your actual needs. Are you looking for some outdoor living space for the family or do you want a big recreational area with all mod cons for entertaining dozens of guests?

Start by writing a list of features you must have and others you'd like to have. If you read this book from cover to cover before you begin, it will give you a good idea of what your options are.

At the same time, consider your own abilities and the amount of spare time you have to take on a large project. Is it realistic to try and complete the heavy work by yourself, or will it be wiser to employ a brickie for the basic work and add the finishing touches yourself? It's easy to be caught up with initial enthusiasm only to lose it as the weekend work continues.

Barbecue courtesy of Barbeques Galore.

Ten Quick Questions

Design and planning demand that you consider what you want from your barbecue area before you begin to build it. Here's a list of questions to keep in mind as you read this book:

☐ How big a barbecue do you need?
☐ How big should the paved area be?
☐ How much workspace do you need?
☐ Is the barbecue for family use or entertaining?
☐ How much can you afford to spend?
☐ Who will do the job?
☐ Will you use brick, slate or stone?
☐ What type of fuel will you use?
☐ Is shelter or shade required?
☐ Will the neighbours object?

Counting the Cost

With every home improvement, cost is an essential consideration. No matter what sort of barbecue unit you decide on, built-in or bought, the overall cost of paving or decking, constructing a pergola, installing special lighting and landscaping will be about the same.

Once you have drawn up your plans and have all the right measurements, it is relatively easy to cost the job. You will know how big an area you plan to pave, whether or not you need a pergola for shade or shelter and whether any timber for decking will be required and how much landscaping will be involved.

The barbecue unit itself will possibly work out to be the least expensive individual item, depending on your design and the area involved.

The overall project could cost anywhere from a few hundred dollars to a couple of thousand. But if you are doing the job yourself, you can spread your actual payments out over a period of time.

To Buy or Build?

Every home owner will come up with a different answer to the question: buy or build. The convenience of simply going and buying a barbecue may be outweighed by where to store it. The big advantage with a small portable barbecue is that you can take it with you on picnics and holidays.

In most cases, the cost of a built-in will be similar to that of an equivalent sized, good quality, portable unit.

Portable barbecues range in price from under $100 for the most basic open grill firebowl type to well over $500 for the top-of-the-range trolley style barbecue with all the accessories.

Some of the simple designs of brick barbecues using solid fuel can be built for around $100 while adding a drop-in style, gas-fired cooktop will bring that to about $250.

It will probably come down to a matter of aesthetics in the end. In this area, the home-built, brick barbecue often comes out the winner because you can design it to suit your garden.

Whether you choose to build a brick barbecue from scratch, or decide to buy a portable unit, your overall budget will be roughly the same. But, like buying most things for the home, it pays to shop around to make sure you have seen a wide range of what's available and checked out how prices compare.

Below: This stone barbecue is beautifully integrated with the stone paving.

Right: This portable barbecue can be wheeled around easily.

Barbecue courtesy of Barbeques Galore.

Which Fuel?

One of the first decisions you have to make is 'which fuel'. The choice between wood, charcoal, special barbecue fuels such as Willow Heat Beads, gas and electricity will affect the whole design. There are arguments for and against each type and of course personal preference is also important.

Wood

Traditionally used for cooking in the great outdoors, wood is not the best fuel for barbecuing because it burns with a flame. Barbecue cooking needs heat — not flames. One flare-up from fat dripping can burn the dinner. It is essential to let flames die down leaving a bed of red hot embers before cooking.

But a wood-fired barbecue is simple to build and operate, requires virtually no maintenance and is the obvious choice if you have access to a supply of free firewood. Once the basics of lighting and attending the fire are mastered, it will provide years of trouble-free service. It is best used with open barbecues and a hotplate.

Charcoal

Charcoal is essentially wood which has already been burnt down. This greatly speeds up the process of turning it into a glowing bed of heat. It is a very effective fuel for barbecuing not only because it provides plenty of heat, but it is ready for cooking in 15–20 minutes, there's no smell and it can be bought in bags from department and hardware stores. Fatty flare-ups are not a problem because you can dampen the fire to control the temperature. It can be used with both open and kettle-style barbecues.

Charcoal lights readily with firestarters. Arrange a layer of charcoal about 5 cm deep on the barbecue placing firestarters strategically on the base at about 15 cm intervals. When they are alight, place more charcoal carefully on top, then let them burn out. This takes about 10–15 minutes and by now the charcoal is ready for you to begin cooking.

Heat beads in a small portable barbecue.

Barbecue Fuel

Special barbecue fuel such as Willow Heat Beads is a convenient fuel, designed for use in portable barbecues. They are a brown coal char product suitable for open grill barbecuing.

Beads are popular for barbecuing because they can be bought in convenient sized bags and provide good heat for relatively long periods of time. Like charcoal, they have less tendency to flare up than wood. They are, however, harder to light than charcoal and take about 30–40 minutes to reach the cooking stage. It is a good idea to use firestarters (solidified kerosene cubes) to get the fire going.

The beads are piled about 7 cm to 10 cm high over half a dozen cubes of firestarters. When they develop a greyish ash, the fire will be ready to cook on. Because beads burn to a fine ash, they tend to fall through a wire grate fire tray. For this reason the hearth should be made from either brick or concrete. Otherwise, you will need a steel plate fire tray.

Left: Starting up a charcoal fire.

Above: This barbecue uses gas supplied from a portable gas bottle which can be filled at most service stations.

Left: This barbecue can use either wood or gas.

If the first attempt to light the fire with these beads fails, you must start again from the beginning. Never, under any circumstances, pour petrol or other flammable liquids on a fire. To do so invites disaster.

Gas-fired Barbecues

Gas is a very economic and efficient fuel, giving the chef complete control over the cooking temperature. You simply turn up or down as required. Of course, if part of the pleasure of barbecuing is lighting the fire, then gas is not for you. But it is the cleanest and most convenient fuel and you can begin to cook almost immediately. The introduction of volcanic rock which gives the food an authentic barbecue flavour has made gas the most popular fuel for portables and built-ins. The volcanic rock is spread out below the cooking surface, absorbs the heat from the burners and glows — just like charcoal or wood embers. This spreads the heat over a large area. The main problem is running out of gas, so it is a good idea to keep a spare bottle handy. Built-in gas barbecues can also be designed for or converted to natural gas.

Electricity

Most electric barbecues available commercially are usually the smaller, portable models with grilling plates. Electricity as a fuel is clean, odour free and easily controllable. Some electric barbecues have lava rock beds, similar to those found in some gas-fired models to mimic cooking over hot coals.

Using an electrically-heated barbecue plate outdoors is difficult and a well-sheltered location is essential. Electric barbecues require a waterproof, exterior grade power point and of course, all electrical work must be done by a licensed electrician.

13

2
DESIGN AND PLANNING

*Creating a Master Plan • The Right Site •
Designing Your Barbecue • Classic Brick
Barbecue Designs • Ordering Materials •
Tools • A Professional Approach*

Your first step is to draw up a master plan of the garden. Don't feel overwhelmed by a totally bare yard that has just been cleared in the wake of builders, an overgrown garden that has been neglected or simply a large, treeless lawn. Time spent at this stage is well spent. Remember the adage: good design does not cost, it pays.

Once you have decided on your design, the next task is to locate the right materials, find the best suppliers and organise deliveries.

For many home owners finding out 'where to get this?' and 'how to do that?' can seem insurmountable. However, the building material supply industry has recently come to realise the vast potential of the do-it-yourself market. Many companies provide helpful product and home handyman instructions in free leaflets and trade literature and give on-going support to building information centres. These centres also conduct courses in many handyman skills for the do-it-yourselfer.

Another useful source of information and cheap building materials is the weekend classifieds. Inexpensive timber and other materials are often advertised in the classifieds where for the price of a telephone call you can shop around for the best deal.

The building industry has one of the largest sections in Telecom's *Yellow Pages*. Through their listings under Building Information Bureaux, Building Materials — Secondhand and Building Supplies, you will find local suppliers of every imaginable building material. Don't assume that the largest supplier or the one closest to you will offer the best deal.

Free delivery seems to be a thing of the past. Therefore, a well-organised builder orders as much as possible from a single supplier, as most companies will charge a flat fee per delivery. It all comes back to effective planning of the job.

Creating a Master Plan

The first and most important step is to gather as much information about the project as possible. A good designer should get to know a client in order to understand his specific needs and taste. Record all relevant data; peculiarities about the site, the prevailing winds, surrounding developments and any future plans for the garden.

Interior designers and landscape architects sometimes suggest that their clients live in a new house for some time, to get to know the special advantages, problems and possibilities of that particular site. In this way, you can avoid many mistakes and identify potential problems.

So, your first step may be to get to know your home better. Whether you have just moved in or have lived there for years, try to look at your house and garden with fresh eyes. Try to think like a stranger who is visiting for the first time.

There are several ways to make this easier. If your house has only recently been built, obtain a copy of the plans from the architect or builder. If this is not possible, remember that some local councils keep microfilm records of plans and may be able, for a small fee, to supply a copy.

Any fee paid to council is good value compared with having to measure the whole house and draw up the plans yourself. You may find that the survey your solicitor had carried out when you purchased the house will show sufficient detail. A survey plan will, at the very least, give you the dimensions of the land and show any improvements.

If all these suggestions fail, it's not too difficult to draw an outline plan of your house and land. Use graph paper (it is sold by most newsagents and stationers) and mark out the lengths of the walls of your house to scale. Now locate trees, fences, existing paths and any other features that may affect your design. Photocopy your site plan so that you can sketch a series of alternative schemes before deciding on the best design. Architects make many preliminary sketches as the development of the design proceeds. Remember, mistakes made on paper are easier to correct than those made in concrete!

When a professional designer completes a set of working drawings, he builds the project, in essence, on paper and irons out all the bugs. This means that when construction time comes, all will be smooth sailing and no unforeseen problems will emerge.

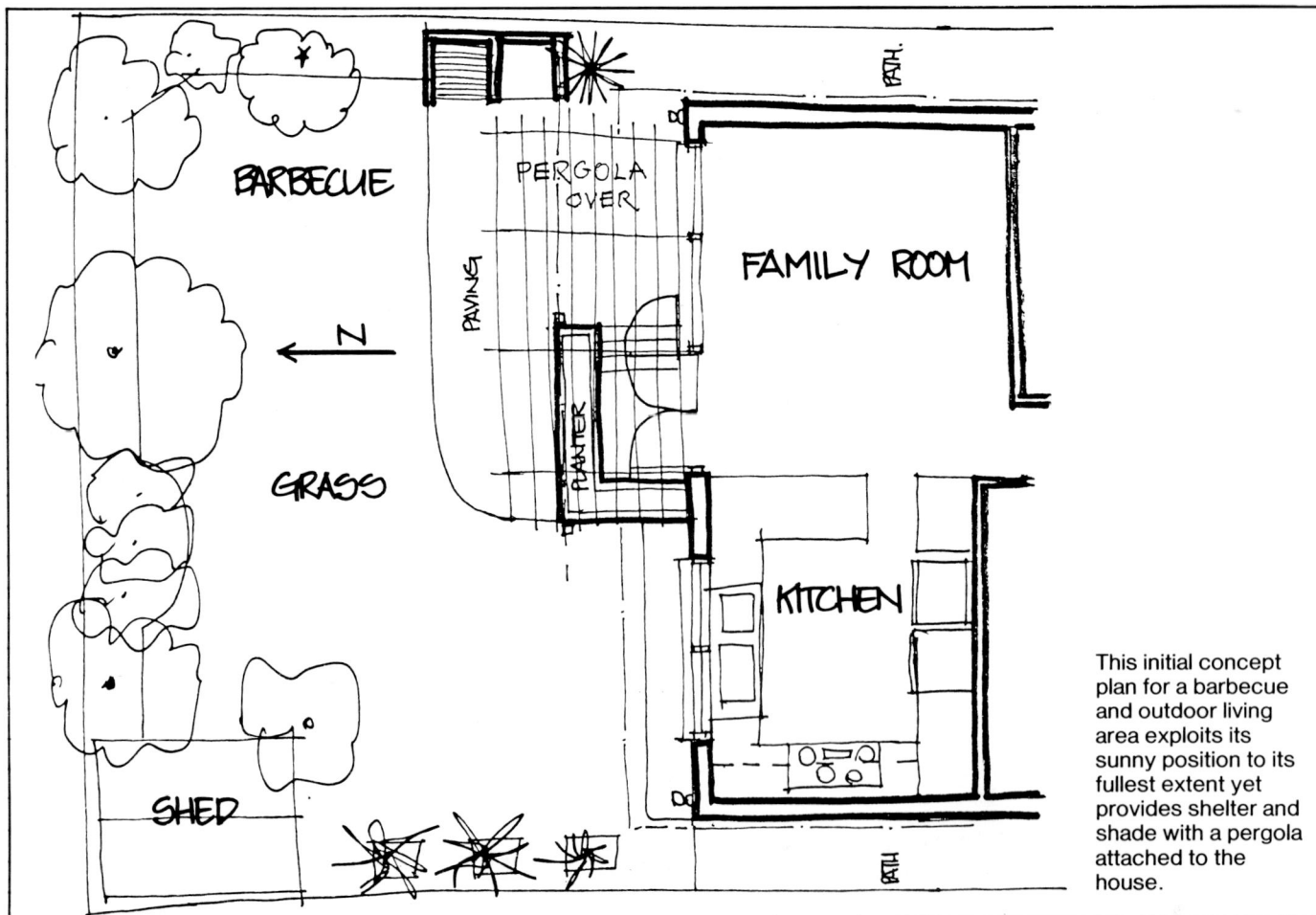

This initial concept plan for a barbecue and outdoor living area exploits its sunny position to its fullest extent yet provides shelter and shade with a pergola attached to the house.

The Right Site

It's always a disappointment to see a well-built barbecue poorly located and sitting alone and forlorn in a forgotten corner of the garden. To create a successful outdoor entertaining area think about:

- ☐ Integration with the garden
- ☐ Shelter and shade
- ☐ Paving, paths and decking
- ☐ Lighting
- ☐ Landscaping

Integration with the Garden

Plan your outdoor entertaining area in the context of your house and garden. First of all the best site is a level site — otherwise your first job will be to level it. If you have a pool, or hope to have one in the future, the entertaining area should be close by for possible pool parties. Because you have to carry food and utensils from kitchen to barbecue, make sure you include a paved pathway with not too many steps. Keep in mind the distance from kitchen to barbecue as well. It is also worth considering the number of people that you will regularly entertain; there's not much fun in being crammed into a corner. Allow plenty of room for people to move around without jostling one another — or the cook.

Keep in mind local council regulations as you may need planning permission in some areas — especially if your building program includes major structural changes like decking or a pergola.

Shelter and Shade

One popular method of linking an outdoor area with the house is to build a pergola. This also provides essential shade for eating outdoors in the middle of the day. A pergola can be covered with a permanent shadecloth or you can plant deciduous climbers for spring colour and summer shade while letting the weaker winter sun through to warm you. Pergolas can also be placed and combined with trellis for shelter from unwanted breezes.

This delightful setting provides the ideal environment for a successful barbecue. Terracotta paving and subtle level changes are used to define the different areas.

Barbecue courtesy of Barbeques Galore.

Above: A serpentine path in brick paving.

Timber decking.

Stretcher bond brick paving.

Exposed aggregate concrete paving.

Paving, Paths and Decking

Paving, paths and decking are an essential component of any outdoor living area. The barbecue and chef need a hard surface which is easy to keep clean and maintain. And guests will soon trample lawn to mud if the eating area is not paved. A deck is an excellent substitute for a paved site area, esecially if your house is built on a sloping site.

Landscaping

The way you landscape the area makes all the difference. Plan your garden to merge in with your barbecue area. Trees and shrubs make for a delightful atmosphere, but leave enough room for the people. Choose plants that will be attractive all year round as barbecues in the cooler months are equally pleasant.

Lighting

Since much entertaining takes place in the cool of the evening, lighting is essential: the chef needs to see the food he or she is preparing and most people prefer to see what they eat.

Permanent lighting should be planned from the very beginning of your design as underground cables may have to be laid. Light is necessary for safety but you can decide whether you want it to be diffused or sharp, inconspicuous or part of the scene. Permanent lights should be installed by an electrician. An option is temporary lighting, usually floodlights using a power supply from the house or atmospheric hurricane lamps or lanterns.

Above: A teardrop light.

Left: Good lighting is essential for barbecue cooking.

Cast iron grills	325 x 475mm	250 x 475mm	200 x 475mm	150 x 475mm
Cast iron plates	375 x 475mm	325 x 475mm	250 x 475mm	a wok plate 325 x 475mm
Angle frames	975 x 475mm	800 x 475mm	650 x 475mm	
Steel plates	800 x 475mm	650 x 475mm		
Steel plates with welded-in angle frames	975 x 475mm	775 x 475mm	675 x 475mm	600 x 475mm
Steel plates with welded-in angle frames and grills	975 x 475mm	675 x 475mm	600 x 600mm	600 x 425mm

A table showing the different measurements
for barbecue components.
As supplied by Barbeques Galore

Designing Your Barbecue

The best place to locate your barbecue is usually at
the edge of the outdoor eating area linked to house
and kitchen and leaving plenty of room for people to
mingle without being jostled. As for the barbecue
itself, there are a number of important design points
to consider.

The biggest mistake most people make is to design
their barbecue before deciding on the hardware which
is available in standard sizes. Visit your nearest bar-
becue specialist first and select the fittings — grills,
hotplates or burners and frames — you will need
before you draw up your final plans. For the actual
barbecue unit, consider the following points:
☐ Size of cooking area
☐ Type of cooktop
☐ Working height
☐ Preparation area
☐ Storage for fuel and accessories
☐ Coping with smoke
☐ Paving

Size of Cooking Area

Your barbecue's size will be determined by the
dimensions of the cooking plates or gas burner you
want and your choice of fuel. If you have chosen a
commercial gas-fired barbecue cooktop, then you
must design the brickwork to accommodate the unit.

The size of the cooking area depends also on the
type and style of entertaining that you intend. For
most families, a plate size of about 900 mm by
600 mm will cope with all but the largest parties. And
even then your barbecue may be supplemented by a
portable unit or a hired one.

A steel plate can be made to suit your exact
requirements if necessary. At larger builders' suppliers and hardware stores you can buy 6 mm thick
steel plate. Some suppliers will even cut it to size for
you. If they are unwilling, a visit to the local
boilermaker's workshop will get the required results.
Don't fall into the trap of buying plate thinner than
6 mm. In the heat of the fire, thinner plates will
eventually warp and become useless.

Type of Cooktop

For the best results with the widest range of foods,
incorporate both a flat steel plate and an open grill
into your design.

Specialist barbecue suppliers have modular cast
steel grilling plates, solid plates and open-cast steel
grills in a range of standard sizes. A visit to one of
these shops is recommended for all intending barbecue builders and designers.

Working Height

A barbecue built too low has two disadvantages. It places strain on the back when long periods are spent cooking and it is less effective in keeping smoke out of the cook's eyes.

It is better to build the barbecue on the high side to help keep the smoke away. The ideal level will depend on the height of the cook but somewhere in the range of 800 mm to 950 mm will suit almost everyone.

Preparation Area

A busy barbecue cook needs to have tongs, spatula, forks, knives, scraper and spoons close at hand. Also needed within arm's length are condiments, some wine and cooking oil for the barbecue plate or grill and a glass of liquid refreshment for the cook. (Cooking is thirsty work.)

A clean, uncluttered space for the food waiting to be cooked is also essential. While it may seem obvious, it is quite unusual to see a barbecue with adequate, built-in working space.

Storage for Fuel and Accessories

A well-designed barbecue will incorporate a good-sized dry storage area for fuel. Wet wood or damp bags of charcoal are difficult to light and cause a smoky fire.

Portable gas cylinders also need protection from the weather. At the outset, make sure your design includes sufficient and appropriate storage for your style of barbecue.

Coping with Smoke

A smoky barbecue is unpleasant for everyone. This problem can be minimised by careful attention to the design and location of the unit. Some of the more elaborate designs may incorporate a chimney.

However, simple barbecues perform well, provided attention is given to siting, wind direction, fences and trees, and the relationship between cooking area, the house and the outdoor eating areas.

A metal hood over the cooktop and the use of good, dry fuels will solve most smoke problems.

Paving

The area immediately in front of the barbecue should be paved to provide a good, dry working surface for the cook. Lawn is unsatisfactory; if the barbecue is used frequently, it will be difficult to keep the grass in good order.

This elaborate barbecue has all the features for successful entertaining: a good working height, with storage underneath; a bright enamel hood to cope with smoke; and, the working area is paved for the comfort of the cook.

Classic Brick Barbecue Designs

In the following pages, you will find several tried and tested designs for brick barbecues that you can build yourself. There is also a materials list which will help you obtain an estimate of the cost for each design.

There is no reason why you shouldn't tackle the job of building your own barbecue. However, bricklaying and concreting are heavy work and require specialist tools. Bricklaying in fact is a highly skilled art and is absolutely back-breaking labour, especially on your own. Should you consider the job too big, there are many bricklayers who advertise 'no job is too small'.

You may decide to hire a bricklayer for the basic structure and complete the finishing touches yourself. That way you can be assured of a professional result with minimum cost.

First, you must purchase the cooktop or burner unit. Whether you are going to build a wood-fired barbecue or one fuelled by charcoal, solid barbecue beads or gas, there is no way that you can be sure that the device you saw in the shop will fit the base you made for it until it is on site.

If you have the unit (it may be a simple steel plate or an elaborate multi-burner gas barbecue) then you will be able to constantly check the fit and make the necessary adjustments while the mortar is still wet.

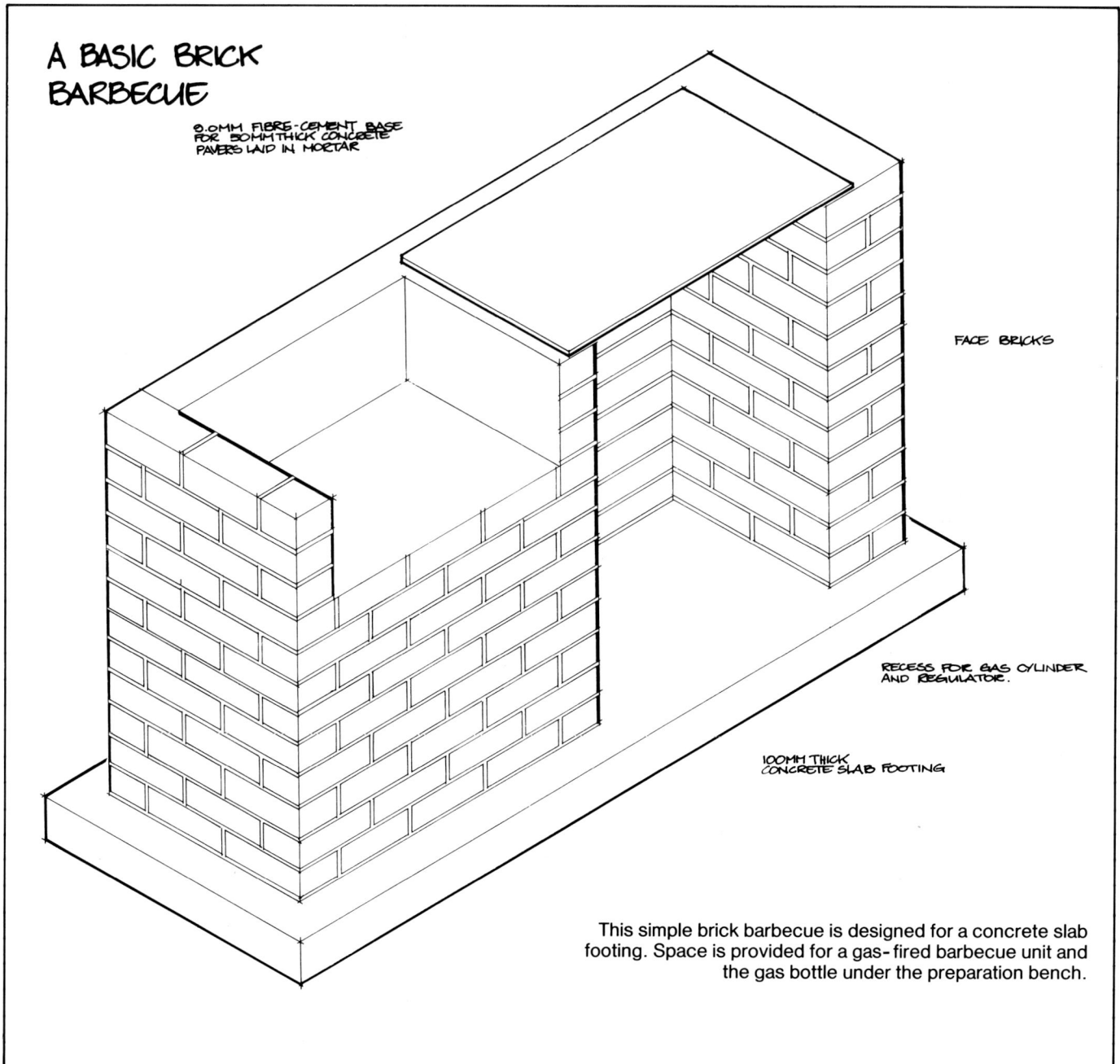

A BASIC BRICK BARBECUE

9.0MM FIBRE-CEMENT BASE FOR 50MM THICK CONCRETE PAVERS LAID IN MORTAR

FACE BRICKS

RECESS FOR GAS CYLINDER AND REGULATOR.

100MM THICK CONCRETE SLAB FOOTING

This simple brick barbecue is designed for a concrete slab footing. Space is provided for a gas-fired barbecue unit and the gas bottle under the preparation bench.

Left: A basic brick barbecue under a shady pergola.

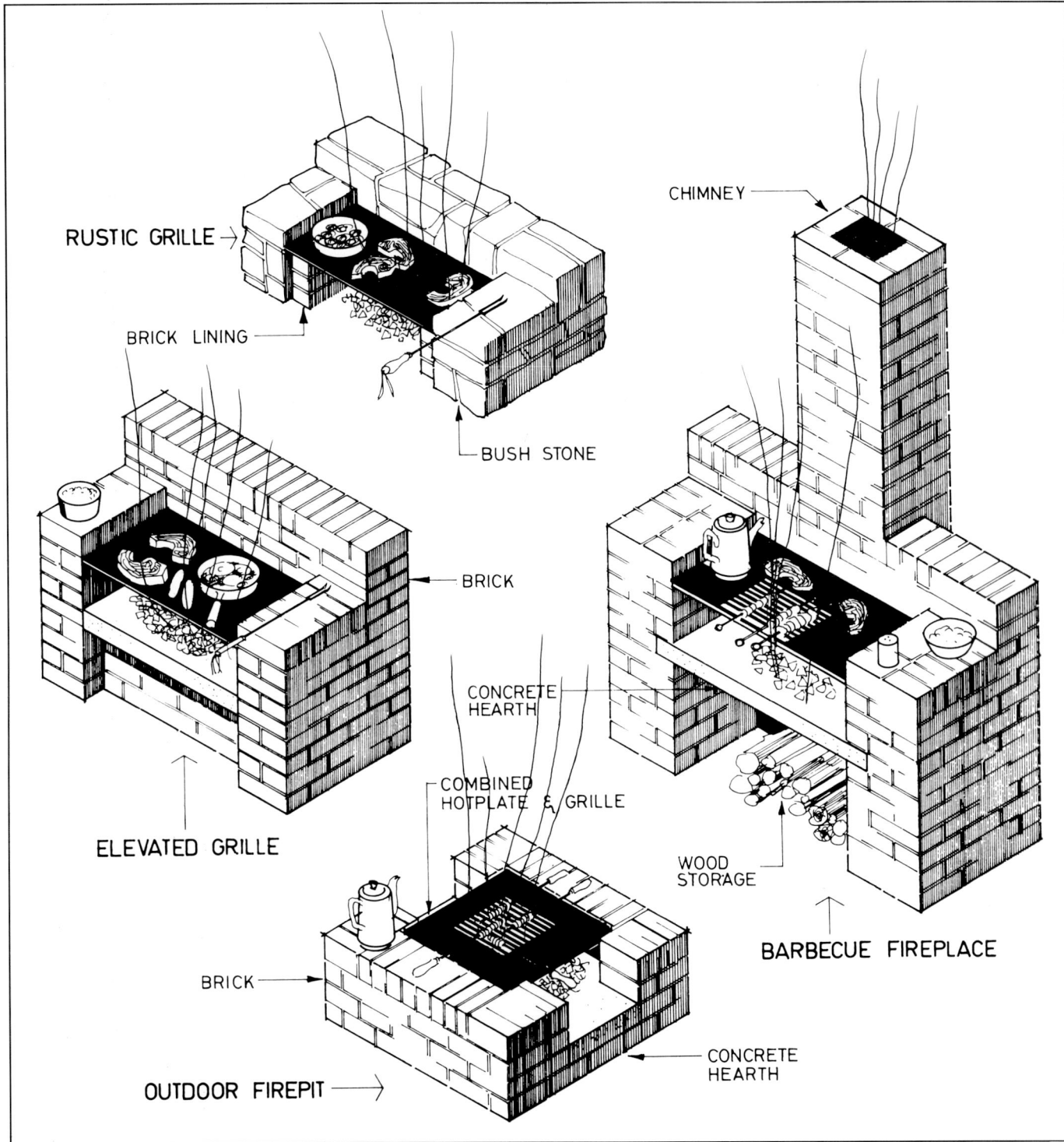

RUSTIC GRILLE →

BRICK LINING

BUSH STONE

BRICK

ELEVATED GRILLE

CONCRETE HEARTH

COMBINED HOTPLATE & GRILLE

BRICK

OUTDOOR FIREPIT →

CONCRETE HEARTH

CHIMNEY

WOOD STORAGE

BARBECUE FIREPLACE

Four different designs of barbecue to suit most locations. They range from a low informal rustic grill to an elaborate design with all the trimmings.

Right: An exploded construction drawing of a substantial brick barbecue.

550

470

240

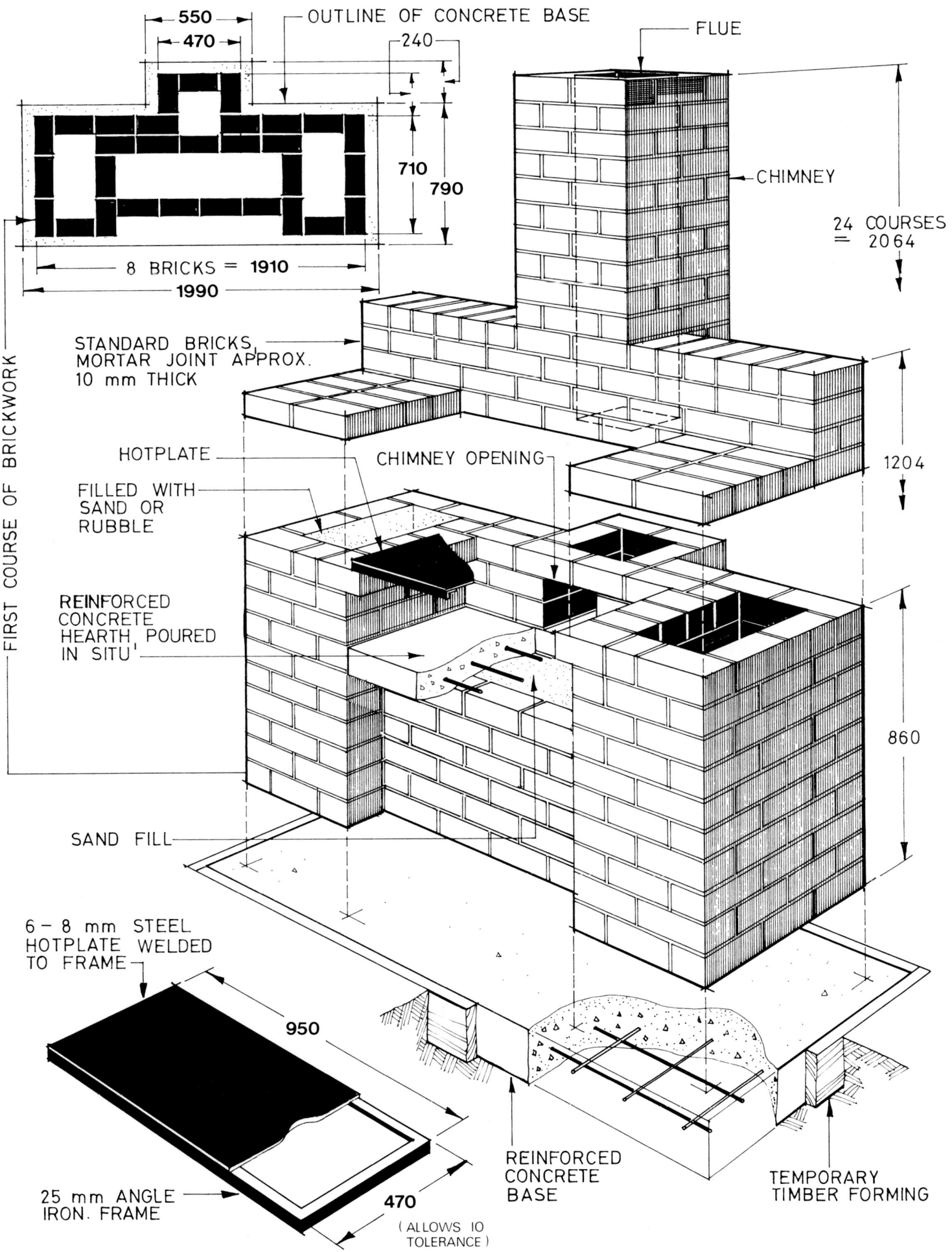

OUTLINE OF CONCRETE BASE

710

790

8 BRICKS = 1910

1990

FLUE

CHIMNEY

24 COURSES
= 2064

STANDARD BRICKS,
MORTAR JOINT APPROX.
10 mm THICK

1204

HOTPLATE

CHIMNEY OPENING

FILLED WITH
SAND OR
RUBBLE

REINFORCED
CONCRETE
HEARTH, POURED
IN SITU'

860

SAND FILL

FIRST COURSE OF BRICKWORK

6 – 8 mm STEEL
HOTPLATE WELDED
TO FRAME

950

25 mm ANGLE
IRON. FRAME

470

(ALLOWS 10
TOLERANCE)

REINFORCED
CONCRETE
BASE

TEMPORARY
TIMBER FORMING

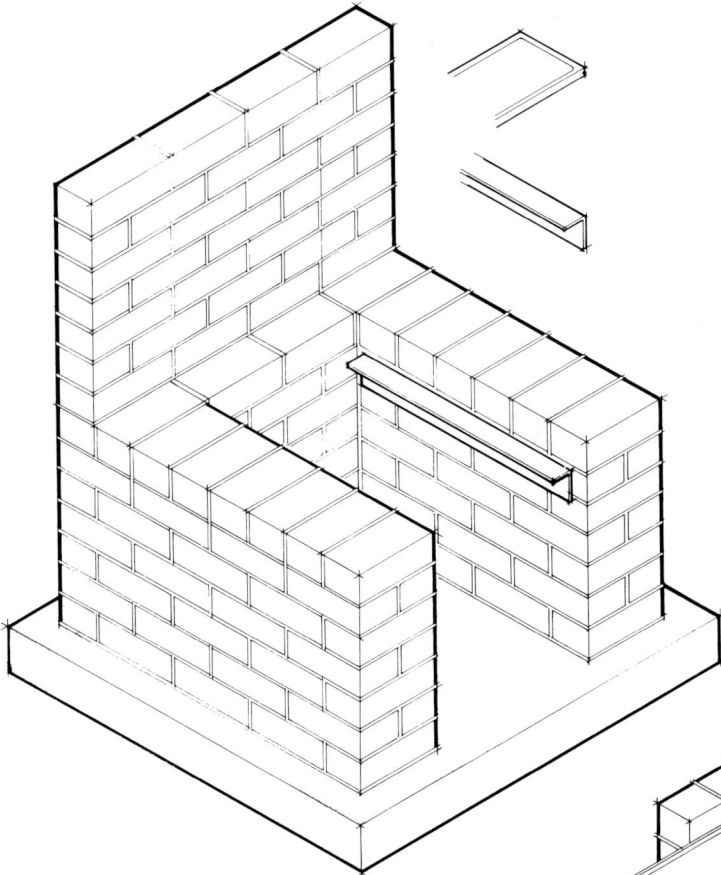

Left: This barbecue is designed for that tight spot where space is at a premium. It is only four bricks wide (950mm) by the same dimension deep. The high back will help keep the smoke from the cook's eyes, while the double brick sides will give space for utensils. About 160 bricks are needed for this barbecue as well as about 0.25 cu m of concrete for the footing. Use a metal fire grate to keep the fire off the ground.

Use standard modular barbecue plates and grill castings for the cooking surface supported on 38mm × 38mm × 6mm steel angles secured to the brickwork with 10mm diameter dynabolts or similar.

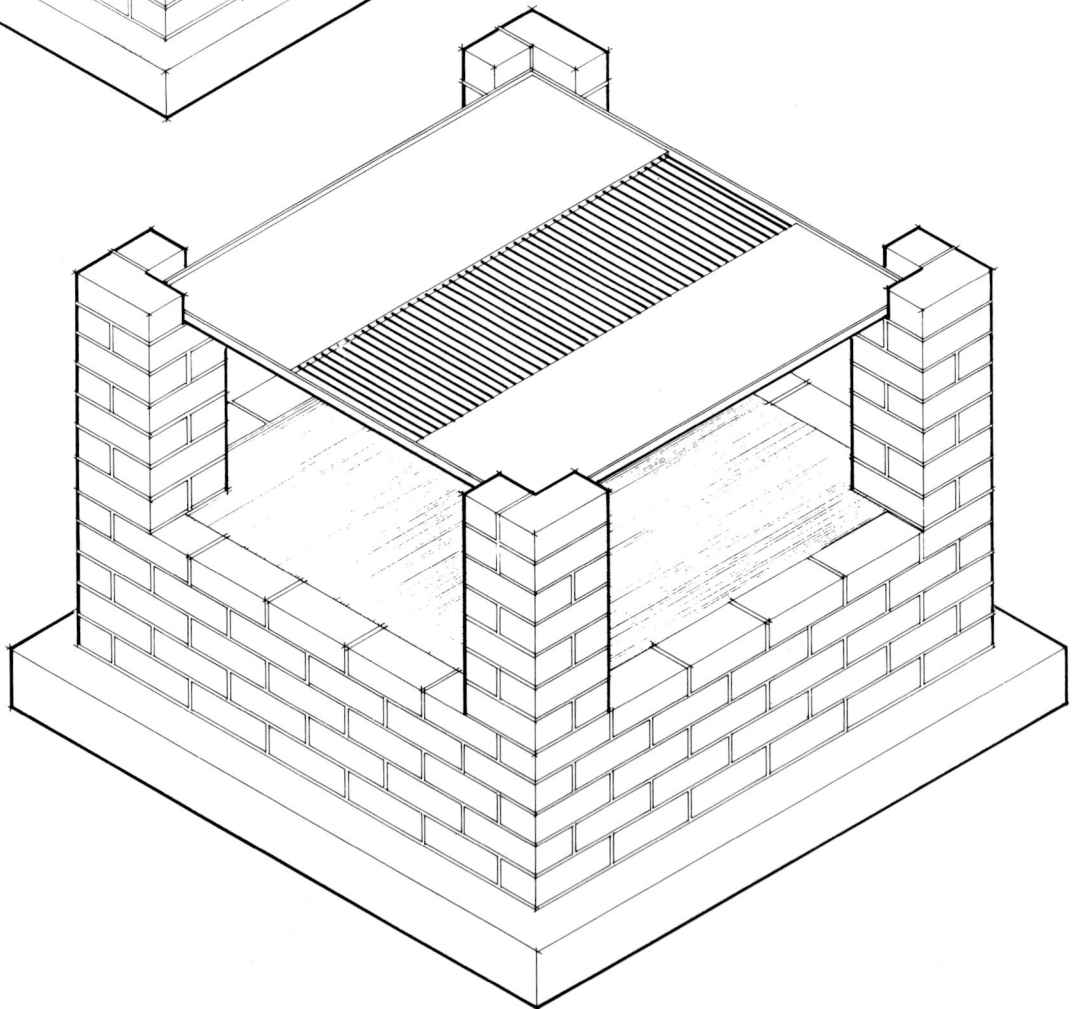

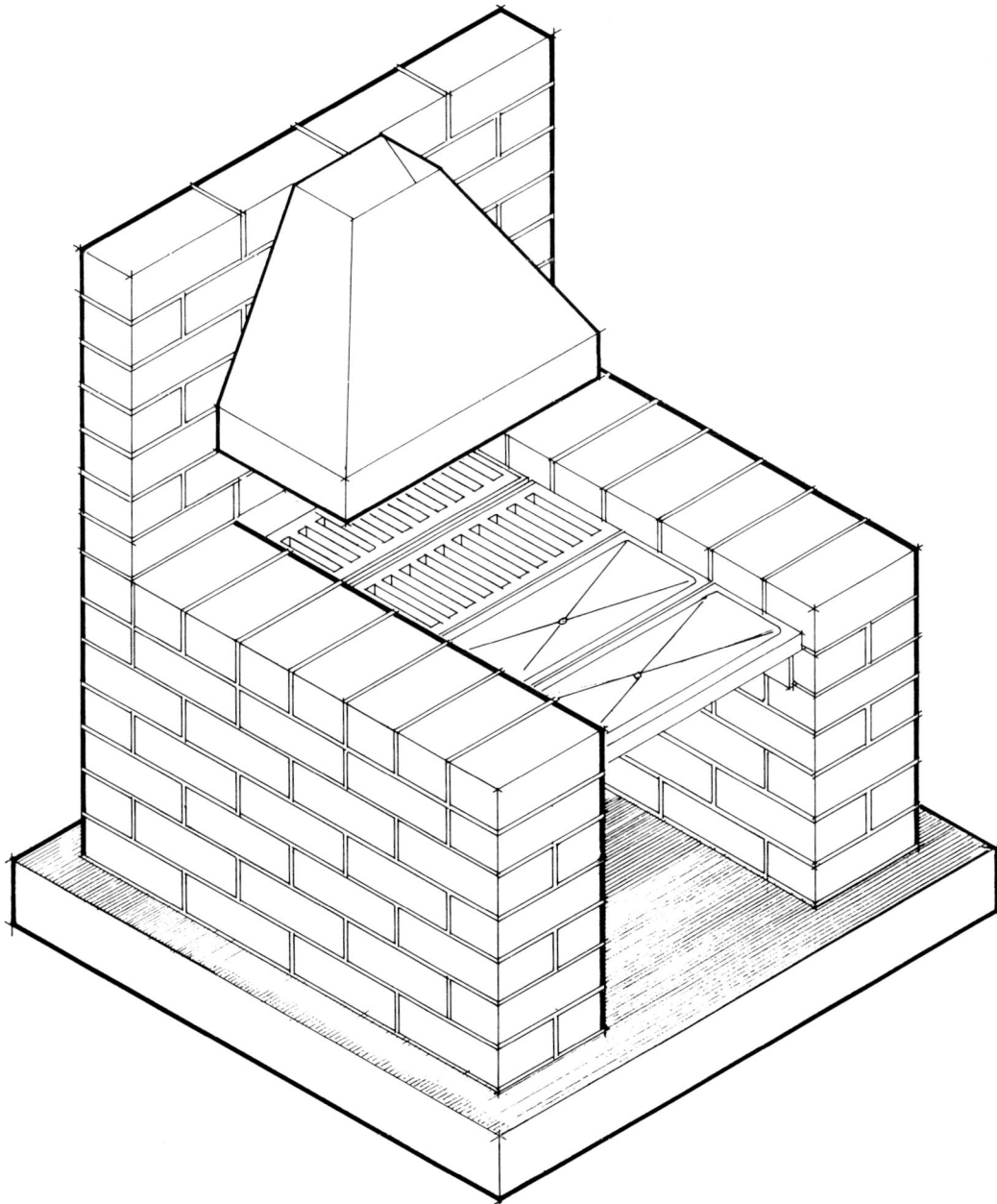

Left: This free-standing design will be a centrepiece of the outdoor living area. The specially fabricated grill plate combination sits in recesses in the top of the four corner piers. A brick hearth is cast inside the brick dwarf wall, while the barbecue rests on a substantial concrete footing. About 250 bricks are needed for this design.

Above: This barbecue has a metal hood for a flue. Use either beaten copper or galvanised steel. A sheet metal worker will be able to give you a quote for the fabrication of your own flue design.

PERGOLA

SERVING COUNTER

PLANTER BOX

WOOD
STORAGE

PAVING

TABLE

SEAT

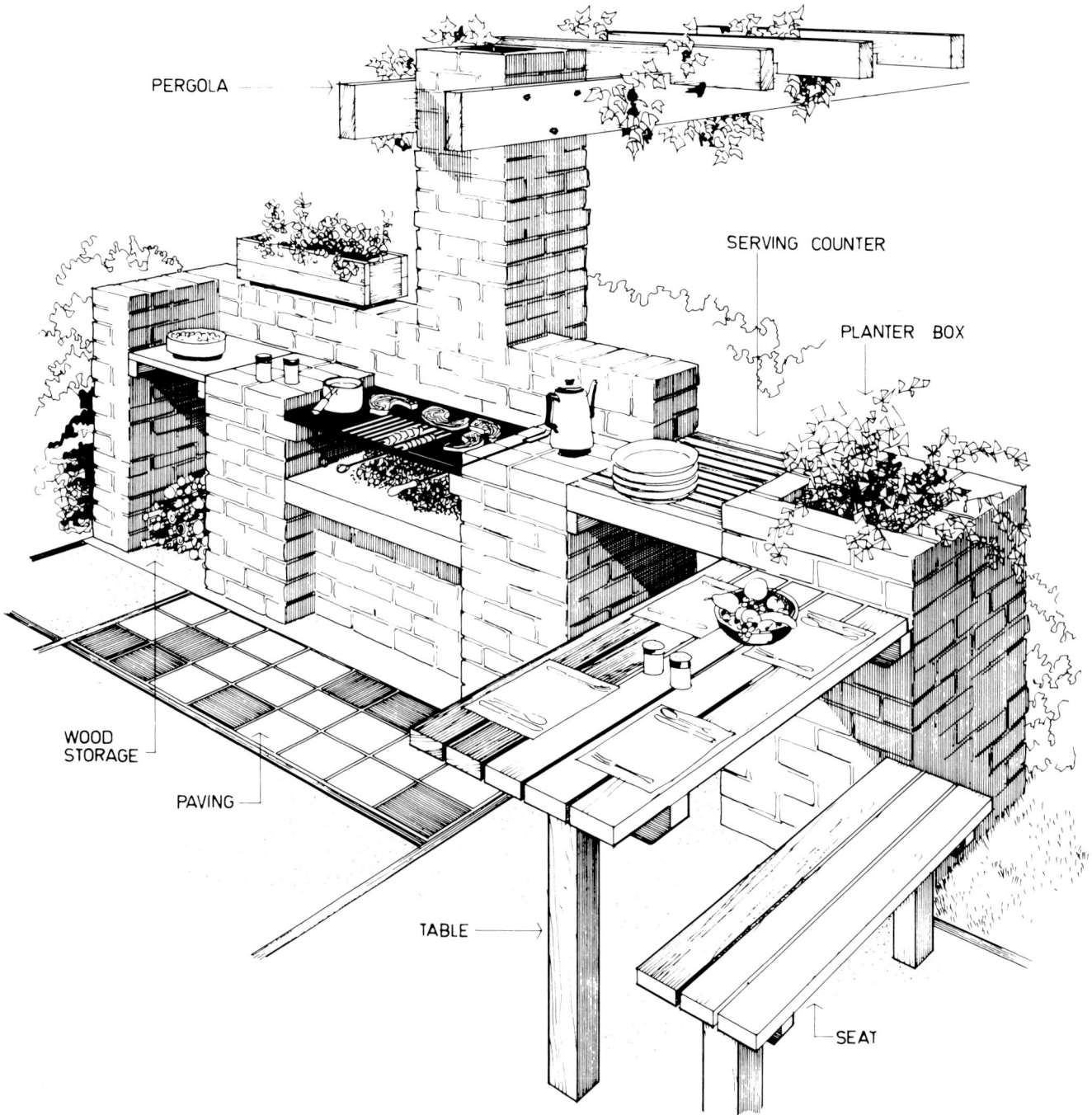

The barbecue that has everything: generous storage for wood, well-designed preparation areas and a built-in table and bench well separated from the cooking area. The barbecue has a flue to keep the smoke away and a pergola overhead to provide shade. Trailing vines around the area give it the final touch.

Ordering Materials

A professional builder will take considerable time to order materials necessary for a job so that maximum discounts are given. Often, discounts are negotiated on the value of an order. But, more importantly, the order should ensure minimal waste.

You not only have to pay a high price for all materials delivered to the job, but what's taken away will also cost you money, time and effort. Take a look at how a project home builder does it. When these houses have been completed, the amount of rubbish left at the end can be taken away with one utility truck.

So, for each job you should complete an exhaustive materials list. It probably sounds elementary but nobody enjoys downing tools on a Sunday (when the job is going well), trying to find a hardware shop open, and paying through the nose for a forgotten item. Your materials list should include all nails and other small items.

If your plans are ambitious, it may be a good idea to open an account at your local builders' supply company. Surprisingly, in these days of high interest rates, a credit account sometimes offers a discount.

Tools

Having the appropriate tools for a specific job is essential. While it is true that good tools will not turn a bad worker into a good one, there are many tasks that are difficult, if not impossible to complete without proper tools.

There is a special pleasure in the ownership of fine tools. Equally, you will need to take special care of your tools for they represent a considerable investment. With reasonable care, good tools should last a lifetime.

There is an old saying amongst tradesmen: there are only two types of tools — good ones and cheap ones. Many do-it-yourselfers learn this expensive lesson very quickly. The temptation to buy a bargain often ends in disappointment as the cheap spade breaks, the hole saw blade dulls after the first hole and the hammer's head falls off. With the job only half done, you may be visiting the hardware shop to replace the cheap tool with the good one you should have bought in the first place.

Most probably, the price of a large tool or piece of equipment such as a cement mixer or a paving vibrator is beyond your resources. However, you can hire items at very reasonable rates. Many hire companies will also deliver the equipment to your home and pick it up when the job's done. Most are also very helpful with instructions in the use of their equipment. If you plan to do a fair amount of work in the garden, ask for a catalogue. It's amazing just what

items you will be able to hire. Check your *Yellow Pages* listings again under Hire — Builders', Contractors' and Handyman's Equipment.

While it may be possible to complete all of the jobs described in this book with hand tools, it is a good idea to have at least a good quality electric power drill with a minimum of 10 mm diameter chuck capacity and a power saw with a blade no smaller than 225 mm in diameter. The saw should have a tungsten carbide-tipped blade as the normal steel blade that comes with most saws will quickly need resharpening. Provided the saw is used with respect, tungsten carbide blades have a life expectancy at least fifty times that of a steel blade.

Power tools not only make the heavy work easier, they help you achieve better results. One of the interesting facts about the use of tools is that it is rarely a sharp tool which causes injury. Rather it is a blunt tool which is being used with excessive force that will slip and injure the user.

Before you begin any job, it's a sensible practice to check that all your tools are in top condition. Also, while you are still fresh and the workplace is tidy, sharpen any cutting tool that isn't up to standard. Nothing is more infuriating than having to climb down from a ladder to sharpen a blunt chisel, or wrecking a joint through persevering with a blunt tool.

A Professional Approach

When professional workers arrive on a job they bring a range of skills and resources not generally available to the amateur. Firstly, they have the experience. They have tackled most tasks before and are aware of the problems that may be encountered in the course of the job.

Secondly, the professional worker has the special tools required for the job and is skilled in their use.

The third advantage of professionals is their access to suppliers, both in terms of availability of materials and better trade discounts.

The tricks of the trade as they relate to the skills of building barbecues and accompanying structures are covered in this book. You probably have no intention of becoming a fully-fledged bricklayer or carpenter but by following the tips outlined in this book, you will be able to get results that will stand comparison with any professional's work.

This book also aims to take away the mystery of tools and techniques so that you will be able to achieve professional results.

It is unlikely that you will get the same treatment as a trade customer. However, by knowing the terminology and some of the tricks of the trade you will be able, by shopping around, to secure a reasonable deal from most suppliers.

3
BUILDING A SIMPLE BARBECUE

Materials • Tools • Site Preparation • Concrete Footings in the Garden • Choosing Bricks • Mortar for Bricklaying • Bricklaying • Building in the Grill and Plates • Finishing Off • Jointing

Because bricks are attractive, economical and easy to handle, brick barbecues are popular with the home handyman. Stone is more difficult to use and tends to be more expensive. But, even if you choose stone to harmonise with your house or garden, the firebox and its surrounds still need to be built of firebricks or well burnt (clinker) bricks.

Although brick and stone laying are jobs normally done by skilled tradesmen, the do-it-yourselfer can achieve very commendable results so long as the design is not too ambitious. It is important that you mark your wall positions carefully with string line and pegs and construct sound footings to provide a level bed for the barbecue. Building a brick barbecue is a good project, even for a beginner, and by doing the job yourself you can vary the pace of the work to suit your budget and social life.

This house, which is also shown on the front cover and pages 66-67, was designed by Douglas McKay.

30

Materials

200 common bricks (depending on design)
2 bags cement
6 bags clean sand
1 wire grille (600 mm x 760 mm)
1 steel plate (700 mm x 350 mm)
2 pieces dressed hardwood, each 2400 mm long (100 mm x 25 mm)
4 (75 mm x 10 mm) masonry anchors (dyna-bolts)

Tools

spade and shovel
bricklayer's trowel
brick set (for cutting bricks)
engineer's hammer
electric drill with 10 mm masonry drill bit

Site Preparation

Clear the site of any vegetation and excavate down to undisturbed soil. In most cases this will be no more than 300 mm below the existing surface level.

As the completed barbecue will weigh around 600 kg, make sure the soil is well compacted to avoid settlement and cracking later. Carefully set up string lines; see page 48 for details.

The base course of any brick wall is called the footing. Some designers prefer a raft footing for a substantial barbecue. This is simply a concrete slab around 100 mm thick and about 100 mm larger than the outside dimensions of the barbecue.

A simple brick footing will suffice for all but the most elaborate brick built barbecues. As a general rule, a brick footing should be one half brick wider than the thickness of the wall it supports. In most cases the barbecue builder uses a single or double brick wall.

There are two main sizes of bricks commonly used. The most common size is called the **metric standard brick** which is 230 mm long × 110 mm wide × 76 mm deep. These dimensions result from the 'soft' conversion of the old standard 9 inches × 4½ inches × 3 inches brick.

The new standard brick has the dimensions of 290 mm × 90 mm × 90 mm. It is known as the **metric modular brick**. The do-it-yourselfer is most likely to use metric standard bricks and for convenience's sake all dimensions are given on the basis that 230 mm × 110 mm bricks are used.

A single brick wall is thus 110 mm wide and a double brick wall has two skins of 110 mm plus a nominal joint of 10 mm, a total thickness of 230 mm. A footing under a 110 mm thick wall is set up with a single brick laid as a header, that is across the line of the wall.

Under a double brick wall, the footing is laid up from one and a half bricks in two courses; the base laid as headers and the second laid in stretcher bond. Some bricklayers prefer to lay their footings with bricks laid on edge.

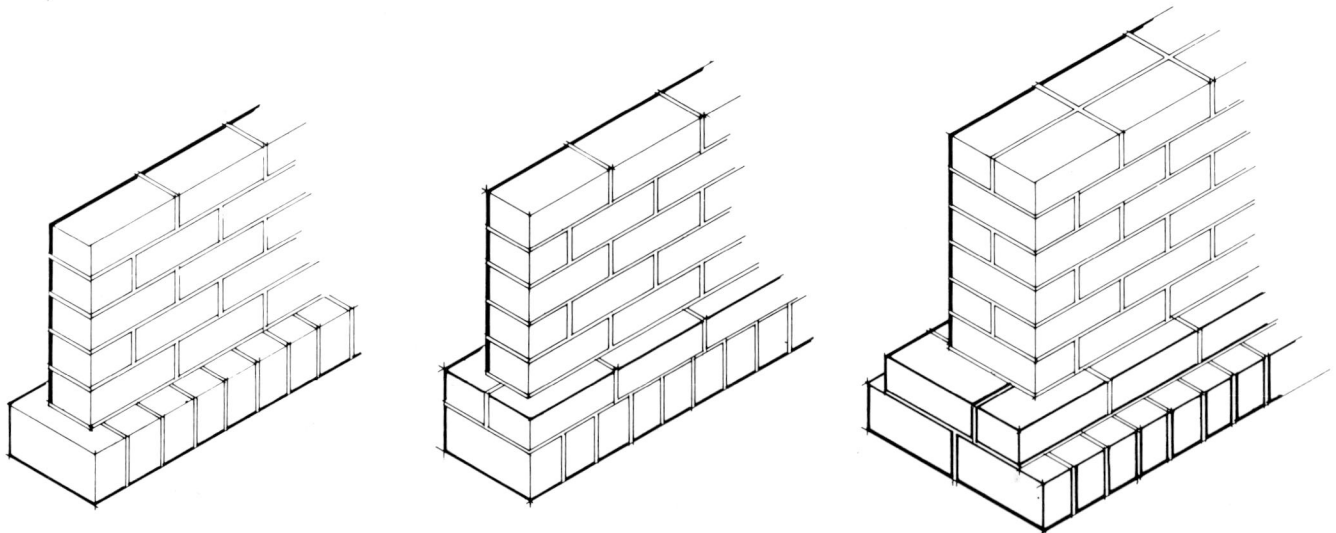

Various types of brick footing for 110mm and 230mm brick walls.

32

Concrete Footings in the Garden

On a reasonably sized barbecue, the footing takes a respectable amount of concrete. A barbecue that's 1800 mm wide by 600 mm deep will be built on a footing some 2000 mm × 750 mm × 100 mm thick. This footing will need about five barrows full of concrete.

Concrete suitable for footings is mixed from stone aggregate, sand and cement in the proportions 4:2:1 by volume. The type of stone used in your area depends on what is quarried locally, but your builders' supplier will be able to provide suitable aggregates.

A concrete footing for the average barbecue is too big a job for manual mixing methods, especially for the amateur. The most sensible way to obtain your concrete is either to hire a suitable electric cement mixer for the day or buy the concrete, ready-mixed from a specialist supplier. Some companies have 'mini-mixers' which will deliver the relatively small amounts of concrete you need for landscaping projects in your garden.

You also need steel mesh reinforcement to make the footing strong enough. Most major building suppliers sell mesh suitable for minor footing works, already cut to your specifications.

Order your reinforced steel so that it is 100 mm smaller than the final size of the footing. The steel mesh is placed so that the cut edges of the mesh are no less than 50 mm inside the finished edge of the concrete. This is called concrete cover and it is essential to ensure that the steel reinforcing mesh is not exposed, nor closer to the surface than 50 mm. Any closer than this minimum and the steel will eventually rust and the concrete begin to spall and lose its strength.

To be effective, the steel mesh should be placed towards the lower, middle section of the thickness of the concrete. This is not as critical for paths and footing work as it is for other exposed slabs but you must ensure that the mesh is not pushed down to the bottom of the slab during the pouring operation.

If you have decided to mix the concrete yourself, it is a good idea to use the 'larry' described below and mix in a bricklayer's barrow. This will save you a lot of bending and lifting as well as double handling of heavy wet concrete.

A sheet of plastic film underlay (Fortecon or similar) under the concrete, prevents the mix from drying out too quickly. Concrete chairs hold the reinforcement steel in the proper position. You can buy both of these at your local builders' supplier.

A timber formwork can be constructed from any strong timber and levelled so that the top edges are at the same position as the required level of the footing pad.

A simple wood float will give the desired finish to the concrete, but take care to make sure that the poured concrete has been well compacted. During the setting time, which may take a few days, keep the concrete damp by lightly hosing down the surface.

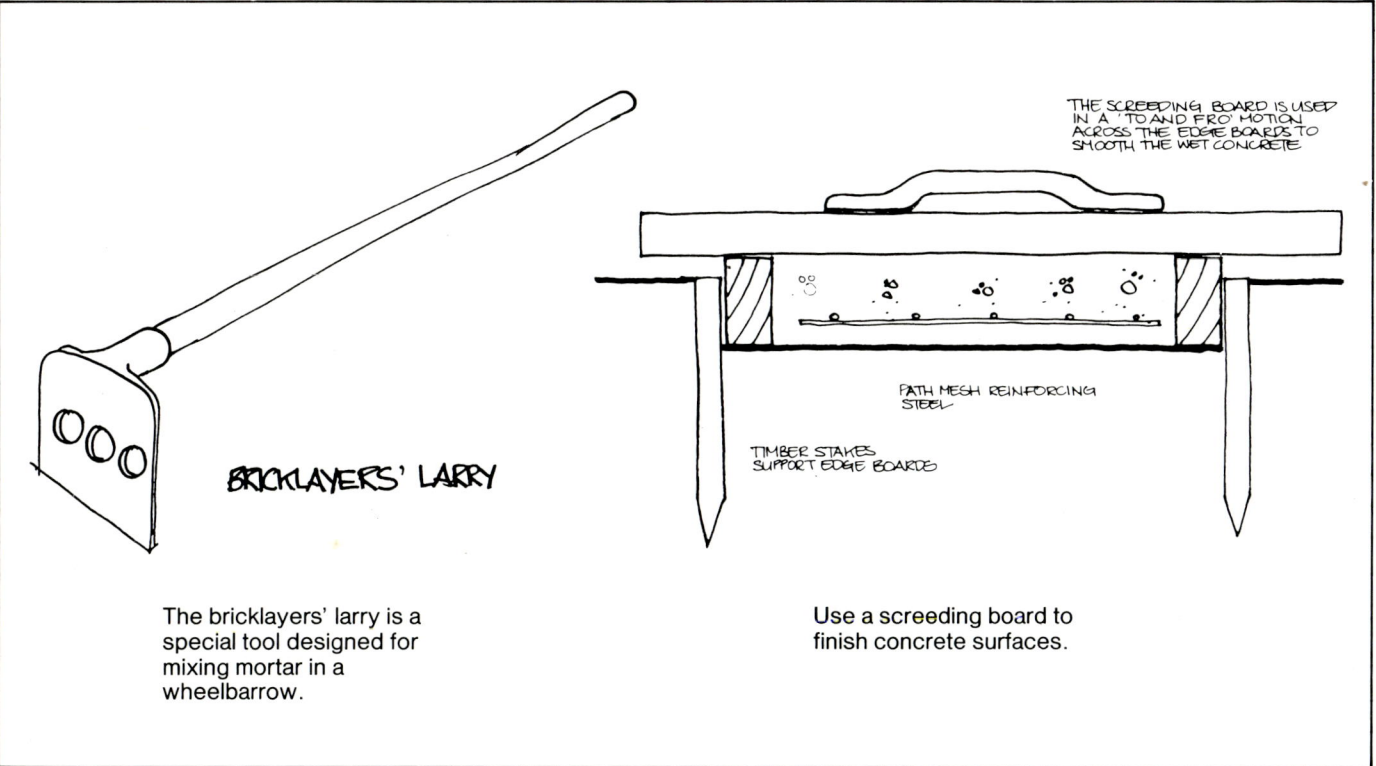

BRICKLAYERS' LARRY

The bricklayers' larry is a special tool designed for mixing mortar in a wheelbarrow.

THE SCREEDING BOARD IS USED IN A 'TO AND FRO' MOTION ACROSS THE EDGE BOARDS TO SMOOTH THE WET CONCRETE

PATH MESH REINFORCING STEEL

TIMBER STAKES SUPPORT EDGE BOARDS

Use a screeding board to finish concrete surfaces.

Choosing Bricks

Selecting bricks is an ordeal well remembered by anyone who has built a new house. The same huge choice of colour, style, texture, material and supplier awaits the barbecue builder.

There are a number of considerations to be taken into account when selecting bricks for a barbecue. Firstly, as a barbecue is going to generate considerable heat, smoke and fat, soft and absorbent bricks will swiftly discolour and become unsightly.

Even the simplest barbecue has a number of exposed corners and top surfaces. With cored style bricks, you will have the problem of covering them up and disguising the holes. Look around and you'll see some of the solutions that have been attempted to fill holes, most unsuccessfully.

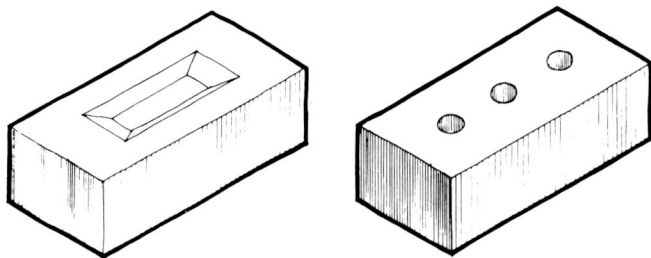

Left: Conventional brick with frog.
Right: A cored brick.

If you have a supply of bricks at hand with cored holes rather than the conventional frogs, then there are two possible solutions. You may design your barbecue to have a tiled capping grouted onto the top surfaces of the walls. Another neat method is to obtain sandstone blocks to end a run of wall. This trick looks really professional.

The preferred brick for barbecue building is a well-burnt common or face brick that has a frog. Second-hand bricks are readily available from specialist suppliers and, if sensibly handled, give excellent results.

Most metropolitan brick suppliers deliver a minimum of about 500 bricks or one pallet lot. Delivery charges can be quite expensive in small lots and buying very small numbers of bricks at the local hardware shop costs the earth. The expense of double handling by the shop is reflected in their selling price. The sensible approach is to plan your purchase so that deliveries are kept to a minimum.

Bricks and other materials required for building barbecues are heavy and the work is hard. The smart worker soon realises that the cost of hiring a good wheelbarrow and a cement mixer for the day is rewarded with a quickly completed job and a reduced medical bill.

The local builders' hire shop is a source of free advice and surprisingly cheap rates. For the do-it-yourselfer, buying a cement mixer is not an economic proposition while hiring costs are quite small. Bricklaying, even for the most simple barbecue, requires an amazing amount of mortar. Mixing mortar on the ground by shovel is slow and back-breaking. Mixing concrete for paths and slabs is madness suitable only for the superfit. In any case, it is impossible to achieve a consistent mix quality, compared with a mixer.

Mortar for Bricklaying

The building trade defines mortar as a mixture of water, sand, cement and lime (plus additives) intended for grouting the joints, usually between bricks. Concrete, on the other hand is a structural material comprising sand, cement, aggregate and reinforcement.

In basic brickwork, bricklayers use two types of mortar mixes — compo mortar and cement mortar. Two types of sand are commonly used and are sometimes combined to give special workability to characteristics. These two types are known as **plasterers'** and **bush** sand.

As a general rule, bush sand is used where a fatty mortar is needed and in compo. Plasterers' sand is white-washed beach sand used in cement mortar and in render.

Using a mixer on even the smallest job results in greater accuracy and better results. If you want to colour the mortar, add your measured quantity of coloured oxides to the mix. Great care is needed so that the end result is consistent.

Most do-it-yourself manuals suggest the use of a gauge box to ensure the ingredients are mixed in the right proportions. When the ratios of cement, lime, sand and aggregate are mentioned, the ratios are by volume not weight. However, using a bucket is just as effective as a gauge box. The bucket should be reserved for measuring and kept dry. With experience, you may find it an advantage to add a handful of cement or lime to achieve the ideal consistency.

Hand mixing mortar is best done on a dry board with a shovel or as the professionals do it, in a barrow with a larry. A bricklayer's larry is a type of long-handled hoe that has several holes in the steel head. Mixing is accomplished quickly and thoroughly by pushing the larry up and down, along the length of the barrow. This method saves a lot of bending and lifting and avoids double-handling heavy mortar.

A quick visit to a building site where a team of bricklayers is working is most instructive. Make sure you don't get in the way and get permission from the foreman on the site before entering private property.

Mix only sufficient mortar for the number of bricks you can lay before the mortar goes off. When cement, sand and water are mixed together, a series of chemical reactions begin. Mortar does not simply dry out, but the materials combine to form the rock-like material we know as cement. For this reason, adding water to partially set (dry) mortar only has the effect of destroying the strength of the joints.

Compo Mortar

Compo Mortar is relatively weak mortar that gives excellent workability. It is perfect for all barbecue work except where higher strength and waterproofness may be preferred below the ground in footings.

Compo mortar is made from cement, hydrated lime and sand mixed to the proportions of 1:1:6 by volume. Only add enough water to give the mix workability. Too much water or too little will result in mortar that is sloppy and weak or dry and crumbly.

Compo mortar is cheaper than full cement mortar and is easier to work with. It will give excellent results for all barbecue work with the possible exception of base courses and foundations where the job will be permanently wet.

Cement Mortar

Cement mortar is commonly used in brick walls and footing work where optimum strength and durability is needed. Cement mortar is mixed from sand and cement in the proportions of 3:1 using similar methods as described in Compo Mortar.

Additives

Mortar additives are a comparatively recent development. Previously, bricklayers would add a handful of lime to cement mortar to make it more workable. They preferred to use compo mortar as it was easier to use. Bush sand as a mixing sand also helps.

The new additives are basically detergents that soften the mixing water and allow the mortar to mix and work more easily. Should you wish to use a workability additive, follow the directions on the can carefully. Resist the temptation to add a little more. Using too much additive means the quality of the mortar, and the strength and appearance of the finished job will suffer.

Bricklaying

Lay out the barbecue in full bricks avoiding, as far as possible, cutting. Your barbecue may be three or four bricks deep by say ten bricks wide. You will have some 25 to 30 bricks per course in such a barbecue.

Although it is difficult to give definite figures, as barrows vary in size and some bricklayers use more mortar than others, a medium wheelbarrow load of mortar will be sufficient for about 30 to 50 bricks. As a typical barbecue has so many corners, it is difficult to lay that many bricks before the mortar goes off. It's better to mix a smaller quantity of mortar more often, which will be easier to mix and will waste less.

BRICK BARBECUE WITH A TIMBER WORK AREA

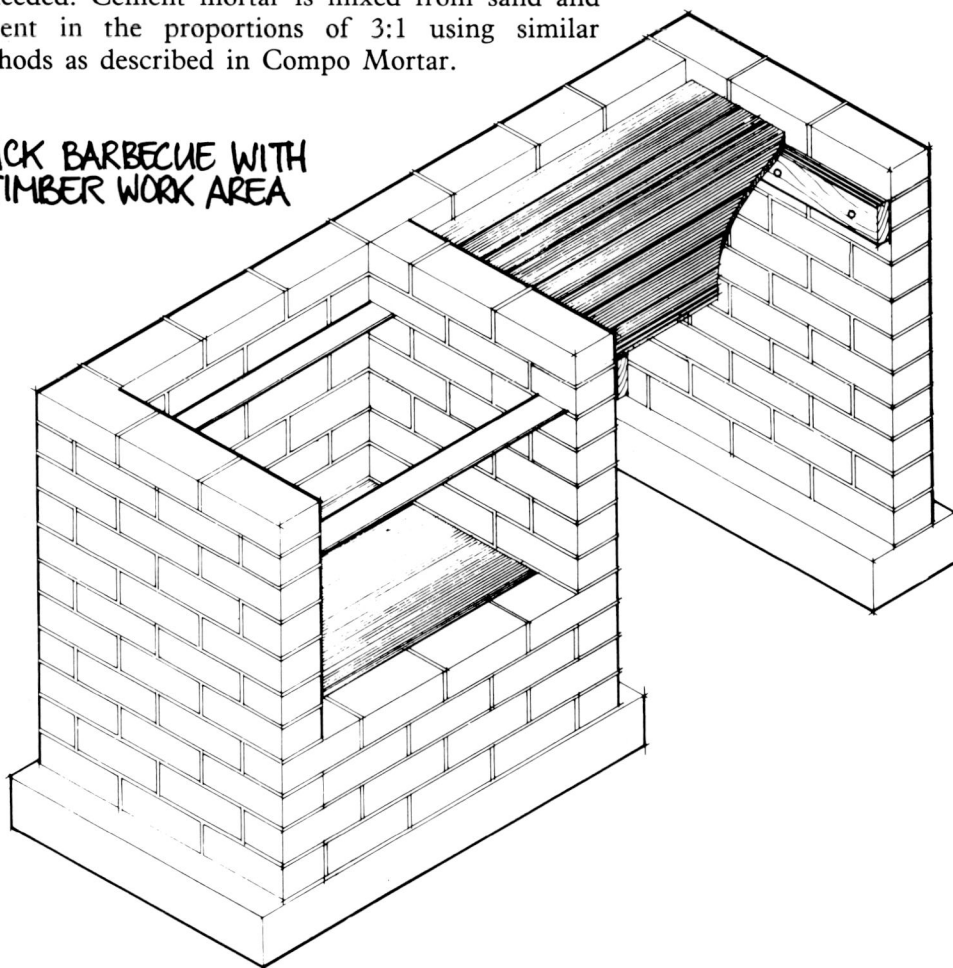

Left: This simple barbecue is easy to build and has everything you need for happy barbecue cooking. It has an ample cooking area, and as the barbecue plates are supported on two flat steel bars front and back you can use any combination of steel plates or open grills.
This barbecue is designed to use the standard 480mm modular steel plates and grills available from Barbecues Galore and other suppliers. About two hundred bricks are needed along with about 0.25 cu m of concrete if a concrete footing is to be used. Allow an extra 25 bricks if you are going to use the simple brick footing methods described elsewhere.

The First Course

Assuming you have completed the concrete or brick footings, you are now ready to begin bricklaying. Mix one full wheelbarrowful of 3:1 mortar and lay a generous strip, at least 100 mm wide by no less than 25 mm thick along the line of the walls. Carve furrows in the mortar with the point of your trowel.

Lay the first brick with the frog down at one corner and accurately position it. Avoid tapping the brick too far into the mortar at this stage. Check that the top is dead level and continue to lay bricks along the string lines until all of the first course is laid. Then, check the level each way and adjust by tapping the bricks into the bedding until the top surface is perfectly level. Time spent at this stage will be rewarded by a quality job. And remember, Don't Hurry.

Second and Succeeding Courses

Once the initial setting out has been done accurately, completing the job is simple, repetitive work. Lay the corner brick in position and set a string line at the edge, away from your working position, to assist with the levelling.

Remember to break bond between courses. This is simply making sure that the vertical joints between the bricks are not continuous. You do this by beginning each second course with a half brick.

Building in the Grill and Plates

The simplest barbecues use small strips of galvanised steel, 100 mm × 25 mm × 6 mm thick, built into the joints to support the plates and grill.

Another solution is to remove some mortar from one joint before it hardens completely and use the recess to slide in the plate etc. If you are going to use a commercial unit for your barbecue, then your bricklaying will have to be designed to fit the system you have at hand.

If the installation of the barbecue plate etc. has been properly considered at the planning stages, then this final stage should cause no problems.

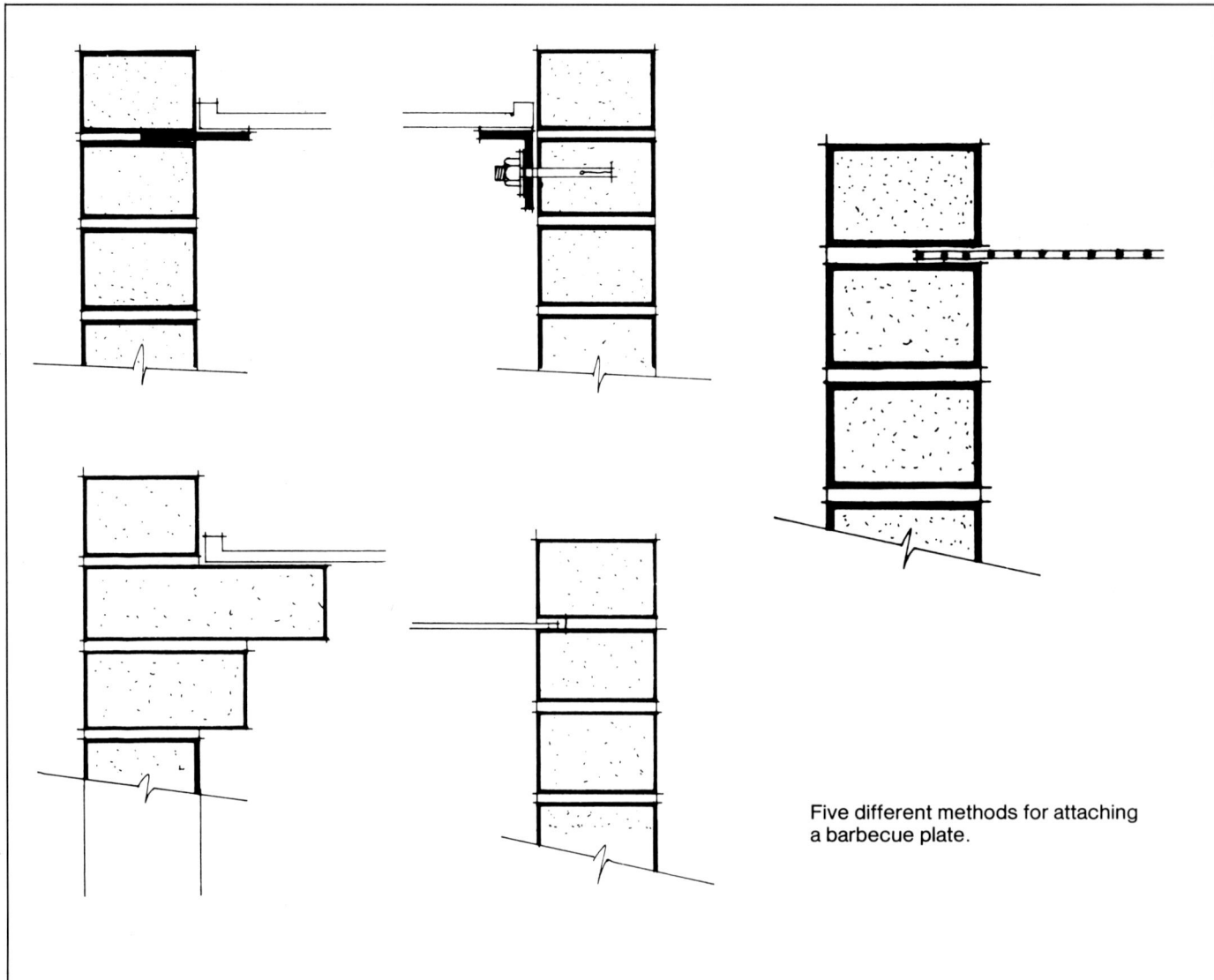

Five different methods for attaching a barbecue plate.

DROP IN STYLE GAS
BARBECUE UNIT

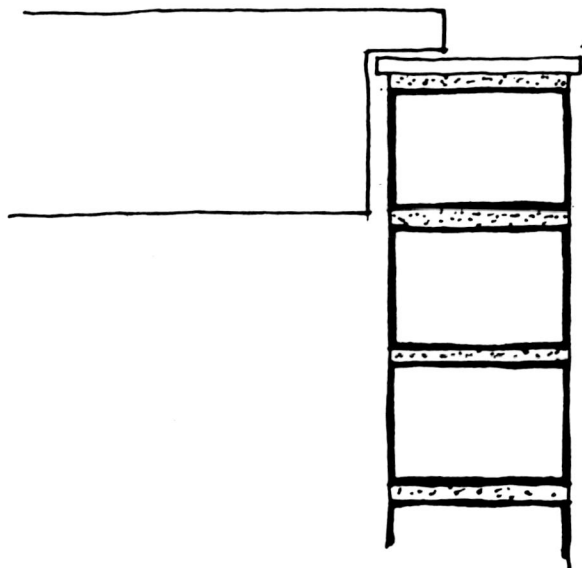

QUARRY TILE CAPPING

SINGLE 110MM
BRICK WALL

SIMPLE MOUNTING FOR GAS BARBECUE UNIT

CAST IRON PLATE AND
GRILL COMBINATION

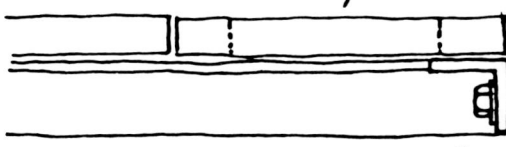

50MM THICK PAVING
BRICK CAPPING TO MATCH.

FABRICATED GALVANISED
STEEL ANGLE FRAME FIXED
TO BRICKWORK.

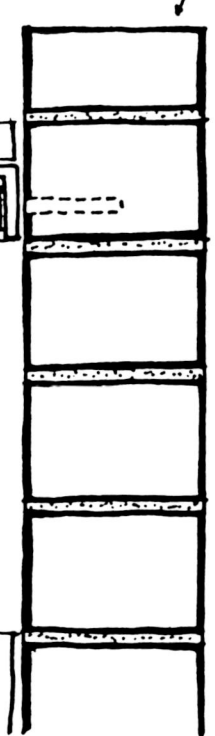

BRICKWORK

BRICK HEARTH

Two different methods of finishing
off a wall. A tiled capping provides an
easy-to-clean surface or you may
use matching paving bricks. Careful
attention to detail will give your
barbecue a really professional finish.

WOOD FIRED BARBECUE GRILL DETAILS.

Finishing Off

Stopping a brick wall when it has reached its designated height may not be particularly pleasing. If you have used cored type bricks rather than bricks with frogs then the problem of disguising the ugly core holes remains.

Installing quarry tiles along the top edge may be an attractive solution. The tiles should be laid in a generous cement mortar, not less than 13 mm thick.

Headers and soldier courses are elegant finishes to the tops of brick walls and suit almost all styles. A header is a brick that is laid across the wall and a soldier course, as the name suggests, is a full brick laid vertically. Cored bricks will still show a core hole at the end walls.

If you have access to a brick saw and have the necessary confidence in your bricklaying ability, a half brick header course, using cut half bricks and laid on their cut ends will provide a neat finish. At the ends of the walls, a mitred cut section closes off the headers and conceals the core holes. This little tradesman's trick provides a really professional touch and it's not too difficult for the amateur bricklayer with a hired brick saw.

Some neat walls are finished off with a slab of specially cut stone. For a really special barbecue where the corner details are prominent, it may be worth the trouble of locating a stone supplier.

Jointing

The joint between the bricks will have to be **struck** and finished in some way. The mortar bed and jointing will be displaced from the joint as the bricks are tapped into position. If you have a large barbecue which is close to your house, the bricks and jointing should match the house.

The simplest joint treatment is to strike the trowel across the joint as the next course is laid and recover the mortar. This leaves the mortar flush with the face of the brickwork with a rough texture. This style of jointing is perfect for all unseen joints and inside the barbecue.

Struck joints are made by running the edge of the trowel along the joint at an angle of about 60 degrees to the line of the brickwork to smooth the mortar and leave a small overhang.

Ironed joints are made with a special tool constructed from a 13 mm diameter steel rod which is 'ironed' along the joint. This gives a smooth semi-circular finish.

Raked joints are done when the mortar is almost hard, usually at the end of the day's work, but you will need to be careful and not let the mortar get too hard. A 100 mm nail set into a flat piece of 75 mm × 25 mm timber at the desired depth (10 mm or less) will serve as a raking tool. The tool is run along the joints, raking out the dry mortar.

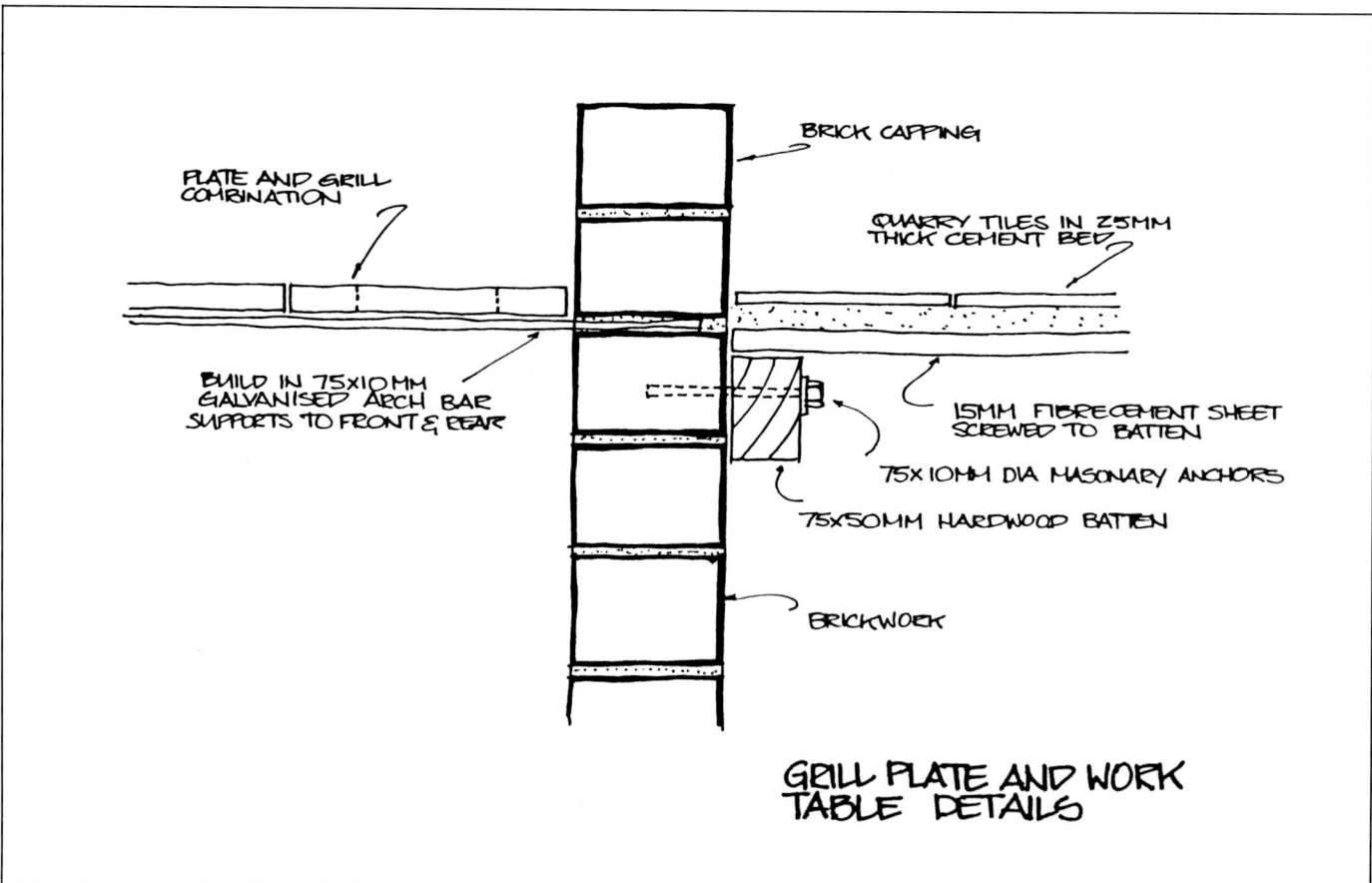

PLATE AND GRILL COMBINATION

BRICK CAPPING

QUARRY TILES IN 25MM THICK CEMENT BED

BUILD IN 75×10 MM GALVANISED ARCH BAR SUPPORTS TO FRONT & REAR

15MM FIBRECEMENT SHEET SCREWED TO BATTEN

75×10MM DIA MASONARY ANCHORS

75×50MM HARDWOOD BATTEN

BRICKWORK

GRILL PLATE AND WORK TABLE DETAILS

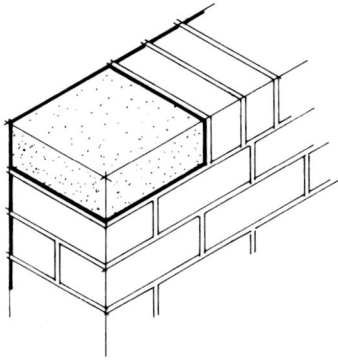

CUT STONE

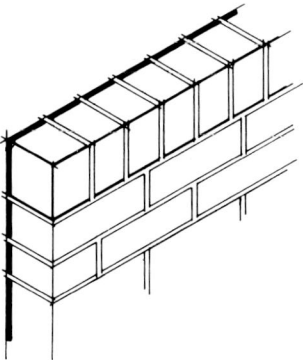

HALF BRICK SOLDIER

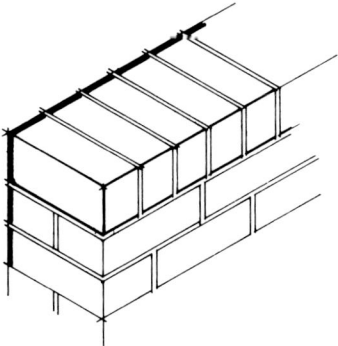

FULL BRICK HEADERS

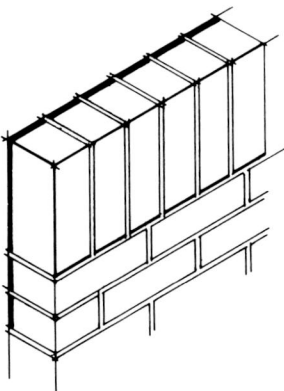

FULL BRICK SOLDIER

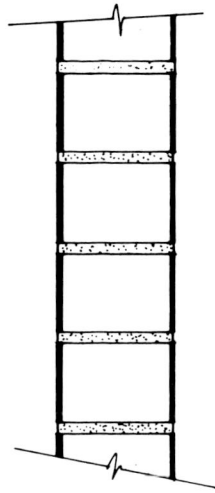

FLUSH JOINTS

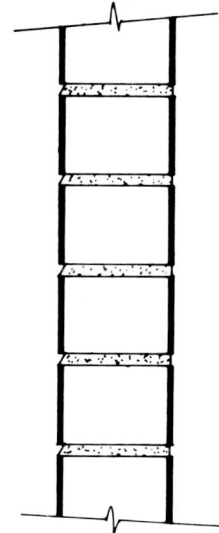

STRUCK JOINTS

RAKED JOINTS

IRONED JOINTS

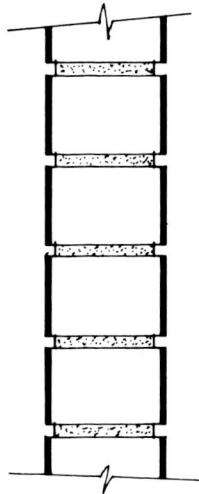

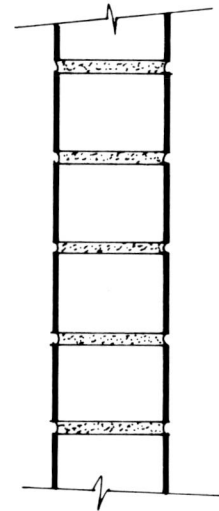

Above: Jointing methods for brickwork.

Left: Finishing off brick walls.

4

AROUND THE BARBECUE: PAVING AND HARD SURFACES

Paving Blocks • Concrete Precast Slabs • Stone and Granite • Bricks for Paving • Paving Styles • Drainage • Site Preparation • Dry Mortar Jointing • Slate and Tiles • Ordering Concrete • Mortar for Tiling • Laying Tiles

Because a hard surface provides a good working area for the cook and is easy to keep clean and maintain, paving is an essential part of any successful project integrating the barbecue into your outdoor living space.

Not only does the area immediately in front of the barbecue need attention, but all outdoor entertaining areas are best paved.

Paving Blocks

The days of boring grey concrete paths and paved areas are past. There is now a fantastic variety of paving materials to choose from. The popularity of brick and concrete unit paving blocks, especially among local councils, is based on some very real advantages. The units can be readily lifted up and replaced as required.

As well as the practical advantages, the huge range of colours, textures, and designs gives you great freedom of choice. Take a look at some of the excellent work being done by local government in malls and public places; this paving work should give you plenty of ideas for your yard.

Common bricks, either new or secondhand, laid in a bed of well-compacted sand or screenings provide an excellent surface for all activities. Block paving is relatively easy, and apart from the basic hard work of handling heavy materials, can be completed by the average do-it-yourselfer.

For paved areas that take a considerable amount of traffic, it might be more sensible to lay the blocks in a bed of mortar than sand. This is especially necessary if the paved area will be used as a driveway for cars.

Existing concrete slabs can be dressed up by the addition of quarry tiles or cobble tiles laid in a bed of cement. First, thoroughly clean off grease and dirt and any other surface treatments from the slab, then

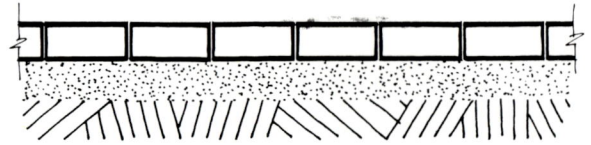

Unit paving is best laid on a bed of sand on screenings.

sparrow pick the surface (that is, use a pick to roughen the surface of the concrete sufficiently) to provide a key for the mortar to adhere to. You can then lay the tiles in the conventional way.

Concrete Precast Slabs

A wide range of decorative, but highly functional precast concrete paving slabs are available from specialist suppliers — just check your *Yellow Pages*.

These may be placed in a similar manner to the paving blocks described above, or introduced as a feature in paved areas which use other materials. Popular finishes include basic concrete as well as a wide range of colours and sizes of exposed aggregates.

As with all paving systems, the most important factor in the success of the project is the time you take to compact the sub-base and sand blinding and the care with which you set out and carry out the job.

Right: Concrete block paving laid in stretcher bond.

Above: Crazy paving in sandstone.

Below: Sandstone paving with cement jointing.

Stone and Granite

Stone flagging is a beautiful surface but may be expensive and time-consuming for the handyman. Surfaces range from sawn sandstone flagging, fieldstone flagging, cobblestone setts, and secondhand demolishers' stone of varying heritage.

Sawn sandstone flagging can be brittle if supplied in inadequate thickness. A minimum of 75 mm thick is recommended although thinner flagging may be laid in a generous mortar bed. Brand new sandstone will range in colour from light yellow through to almost red. Used flagging will probably weather to a grey or bleached white colour.

Occasionally, granite setts are available from demolishers' yards but you will have to make regular visits to obtain the desired amounts as this excellent paving material is much in demand.

Cobblestones are also available from demolishers for those patient builders, patient enough to wait until supplies come to hand. Most demolition yards also have good supplies of foundation sandstone blocks which are quite regular in dimension and can be split to provide a very attractive paving material at a reasonable cost.

Above: Herringbone paving with stack bond edging.

Bricks for Paving

Brick manufacturers have recognised the tremendous growth in popularity of brick and unit paving and have introduced many new products specifically designed for paving.

The normal common brick is less than ideal for paving work. Common bricks have either a frog or are cored to help bonding. The frogs make the task of levelling a little more difficult and, of course, cored bricks are totally unsuitable for paving.

Special bricks for paving are designed to have the length equal to twice the width. This feature allows tight joints in most patterns. Some premium paving bricks also have specially embossed upper surfaces.

Paving bricks are usually thinner than normal bricks, 50 mm thick rather than 76 mm thick. Also, paving bricks are generally well-burnt during manufacture to give superior hardness and are made to closer tolerances than ordinary common bricks. All in all, using these specially designed paving bricks results in a better job, whether done by you or a paving contractor.

Paving Styles

The common brick is 230 mm × 110 mm × 76 mm. These dimensions were designed so that two half bricks plus a 10 mm joint are the same size as a full brick.

In paving work, you will want to avoid this 10 mm joint which is difficult to make accurately and encourages weeds. For paving which uses common bricks, a number of patterns are used to ensure that the joints are tight.

Right: Common bricks used as paving laid in stretcher bond.

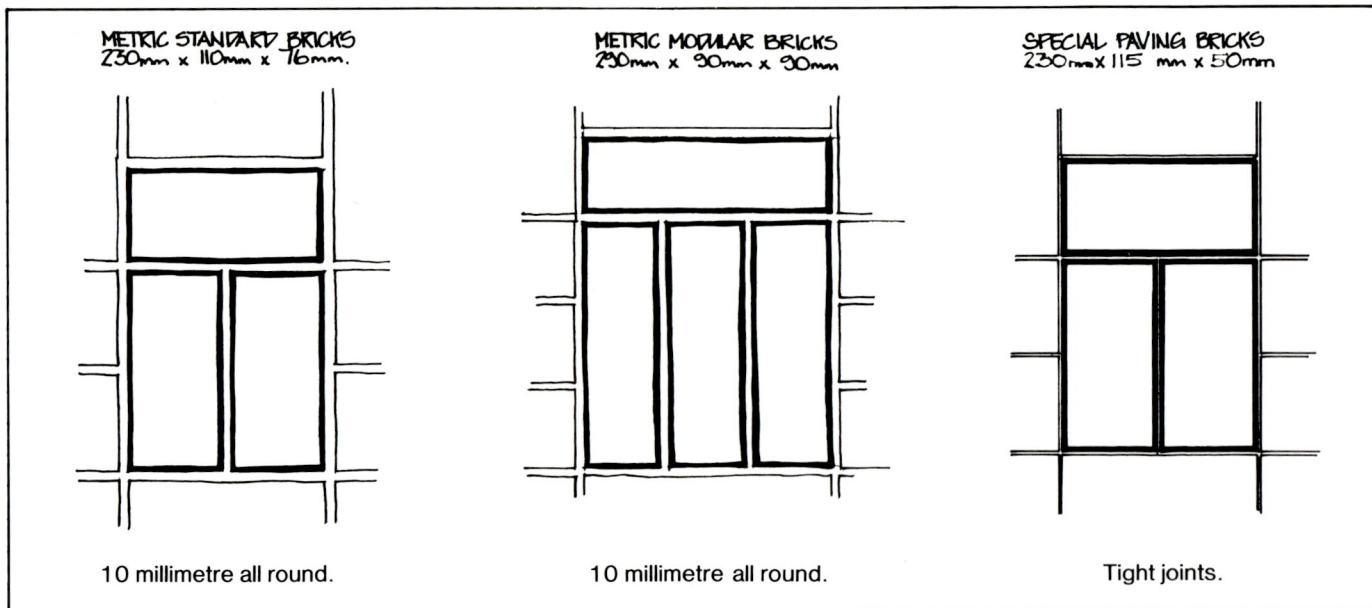

METRIC STANDARD BRICKS
230mm x 110mm x 76mm.

METRIC MODULAR BRICKS
230mm x 90mm x 90mm

SPECIAL PAVING BRICKS
230mm X 115 mm x 50mm

10 millimetre all round.　　　10 millimetre all round.　　　Tight joints.

Stack bond paving in common bricks with borders from timber sleepers.

Stretcher Bond

Stretcher bond joining is the pattern you find in the vast majority of brick walls. Each course of bricks has the short dimension joints displaced by one half brick so that the joints are discontinuous.

Stretcher bond is easy to lay, but requires care to maintain the straight lines. The use of an accurate stringline is recommended. This type of pattern does not require much brick cutting, which is both time-consuming and tedious. Make sure the joints are tightly closed to prevent weeds.

Stack Bond

Stack bond paving is easy to lay and tolerates slight variations in the size of the bricks. Remember that because bricks can vary in size this can cause problems by accumulating errors. Careful work is essential to ensure a neat job.

At the outset, it's important that you follow standard setting out procedures and make the corners perfect right angles. Stack bond is, as the name suggests, a pattern made by stacking the bricks so that the joints form lines in both dimensions. You can vary this by stacking the bricks as before, but breaking bond in each stack, as in the stretcher bond style.

Herringbone

Herringbone is an attractive pattern laid along the line of the borders or on the diagonal. Each brick is laid at 90 degrees to its partner. Along the lines of the work, each succeeding course is displaced one half brick.

Once the initial setting out has been done, this type of paving is quick and uncomplicated to lay. However, laid on the diagonal, you will have to do some cutting. For really professional results, it's best to use a brick saw. Your local builders' hire yard will be able to supply the saw and show you how to use it.

Basketweave

Basketweave uses bricks in pairs, laid alternately 90 degrees to the next pair. If you intend to use common bricks, a 10 mm joint will be required and this may be difficult.

Should you decide to basket weave, it's best to use special paving bricks. These bricks are 230 mm × 115 mm (or other combinations so that the length is exactly equal to twice the width). This lets you close the joints up tight.

46

Circular and other Patterns

There is no limit to the range of patterns for brick and block paving. Circular patterns, fan shapes, and combinations of two or more styles are used to great effect by landscape designers.

Circular paving styles can be used to great effect on a circular courtyard. Careful setting out is essential to keep cutting to a minimum. It is preferable to drive a stake into the ground at the centre of your circle, start from the outside and work inwards.

Use a string line or a radius board to check that your circle is perfectly round. The reward is a truly professional looking job. When you have reached the centre of the circle, there will be a fair bit of cutting and testing required. Sometimes, a centre piece of a different material can be used successfully.

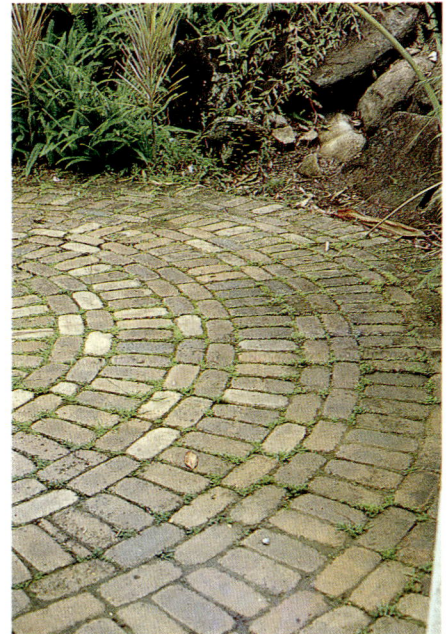

Drainage

An extensive paved area collects a surprising amount of water when it rains. There are ways of making sure a short rain storm does not interrupt a carefully planned barbecue. Carefully provide falls away from the walls of the house and where possible install a sump or gully connected to the stormwater lines of the house.

The new plastic piping systems have accessories that solve all stormwater dispersal problems. Even if the existing stormwater lines cannot be found, it's a simple exercise to construct a dispersal sump at a convenient position in the garden.

At its most basic, a dispersal sump is formed by digging a hole about one cubic metre in size, and filling it with rubble. Use a shadecloth as a filter fabric around the sides of the hole and over the top before covering it over and you will prevent the sump from becoming clogged up with sand and silt.

Lay all drainage lines to falls in 90 mm plastic stormwater pipe so that runoff and stormwater is directed to the sump. Unless it is specifically permitted by your local drainage authority, never connect stormwater or seepage drains into the sewerage sytems.

If possible, the best way to disperse drainage water is through the stormwater system to the street gutters. If you are in any doubt about the best method to direct stormwater away from your property, a visit to your local council's building or engineering departments should provide the answers.

Site Preparation

Setting Out

Setting out the paving is a critical phase that demands concentration and plenty of time. Once you have finalised the design and the materials have been ordered and delivered, you need to indicate positions of all important elements of your plan on the site.

The basic tools for this are a string line, a level and a tape measure. First, locate the corners of the paved area. Builders use sight boards made from two stakes, driven into the ground with a piece of timber nailed across the top of the stakes. The stakes should straddle the string line which will mark the edges of the work. Drive in one stake on each side of the intended stringline and nail in a horizontal piece as a bridge. The nail on which the stringline will be attached may then be located accurately.

The smart way to avoid work later is to set out all paved areas as a multiple of the individual units. In this way, you eliminate almost all cutting. In other words, the area is said to be (say) 50 bricks long by 100 bricks wide. Remember, all bricks are not precisely the same size and common sized bricks are 230 mm long by 110 mm wide.

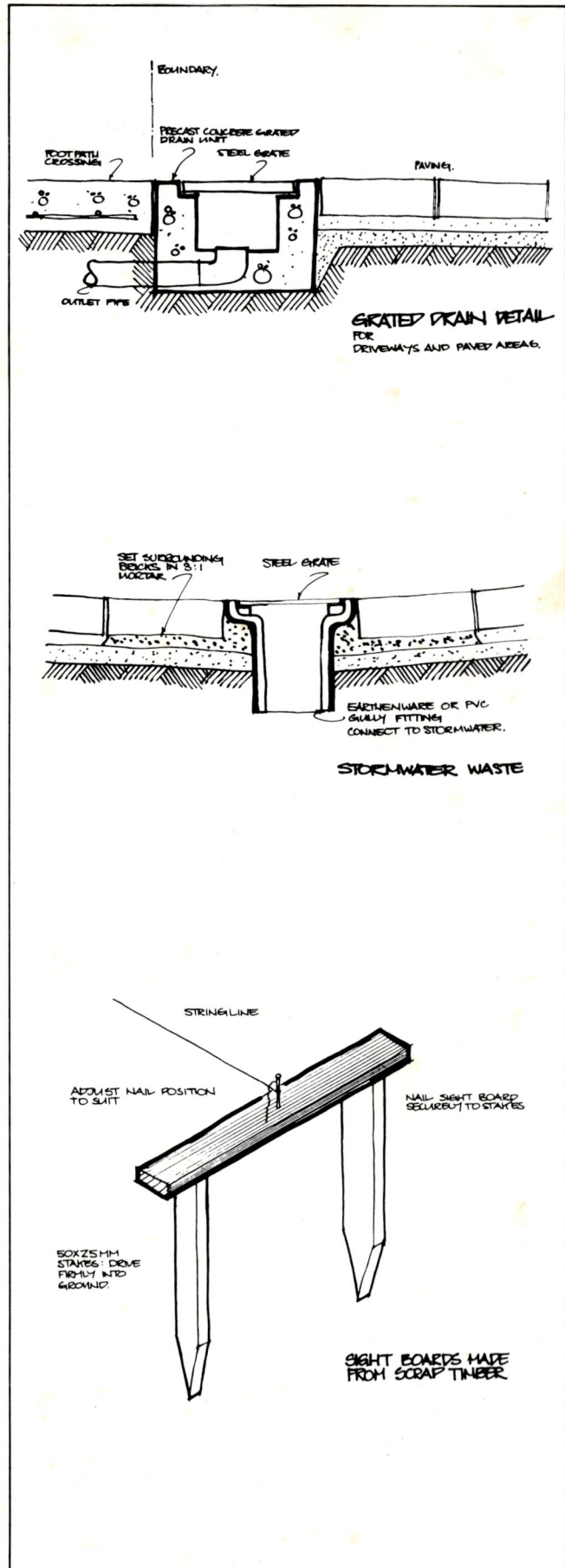

GRATED DRAIN DETAIL FOR DRIVEWAYS AND PAVED AREAS.

STORMWATER WASTE

SIGHT BOARDS MADE FROM SCRAP TIMBER

Removing Organic Matter

If your paved area is replacing lawn, then all grass and root matter has to be removed to give a satisfactory foundation. If you lay paving over plant matter and roots, two problems arise. Firstly, as roots rot and decay, the paving will sag and become uneven. Secondly, roots send shoots through the gaps between the paving units and become a nuisance.

The proper way to start the job is to remove enough soil until you reach undisturbed soil, some 150 mm to 300 mm below. To reduce further problems of plants growing between the gaps, treat the soil with a residual weed killer. Never lay paving over plastic film as this inhibits drainage and results in moss and slime growing in and on the bricks.

The Sub Base

The best sub base for unit paving is screenings — the small slivers and particles of rock left in the crushers of the rock quarries. The material is quite cheap and will not support any root growth. A layer of some 100 mm to 150 mm of screenings provides an excellent base and gives long term drainage to the paved area.

The sub base should be carefully levelled with a screeding board (made from a straight piece of 150 mm × 25 mm timber), then well compacted by ramming.

A thin layer of sand is used as a bed for the bricks or paving units. Ideally this bed should be no thicker than 25 mm. Lay small sections of bricks at a time, constantly checking that the falls are consistent and that the surface is true.

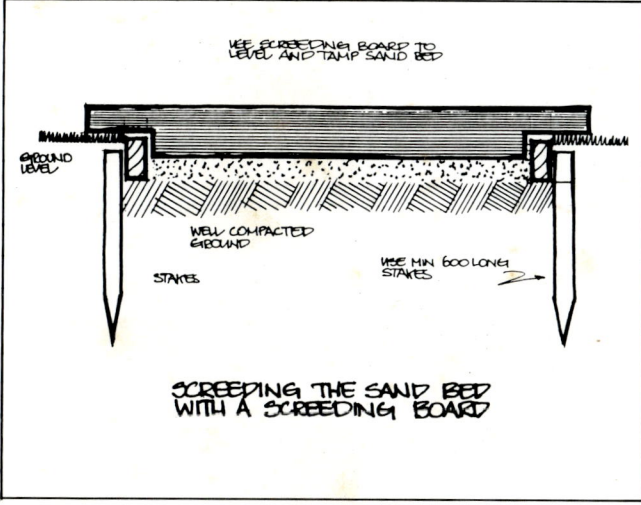

SCREEDING THE SAND BED WITH A SCREEDING BOARD

Final Levelling

Final levelling is done by using a short length of heavy timber and a heavy hammer to tap each unit lightly into position. For larger areas, it is best to use a mechanical compactor. Many hours of back-breaking work can be saved for the cost of a single day's hire of such a machine.

Perimeter Paving

If your paving is likely to receive heavy foot traffic, it is a good idea to lay the perimeter paving in a cement mortar rather than in sand. Using a contrasting border, firmly secured in place, is also a good idea. Treated hardwood, wooden sleepers and bricks and blocks laid as a header course, also make effective borders.

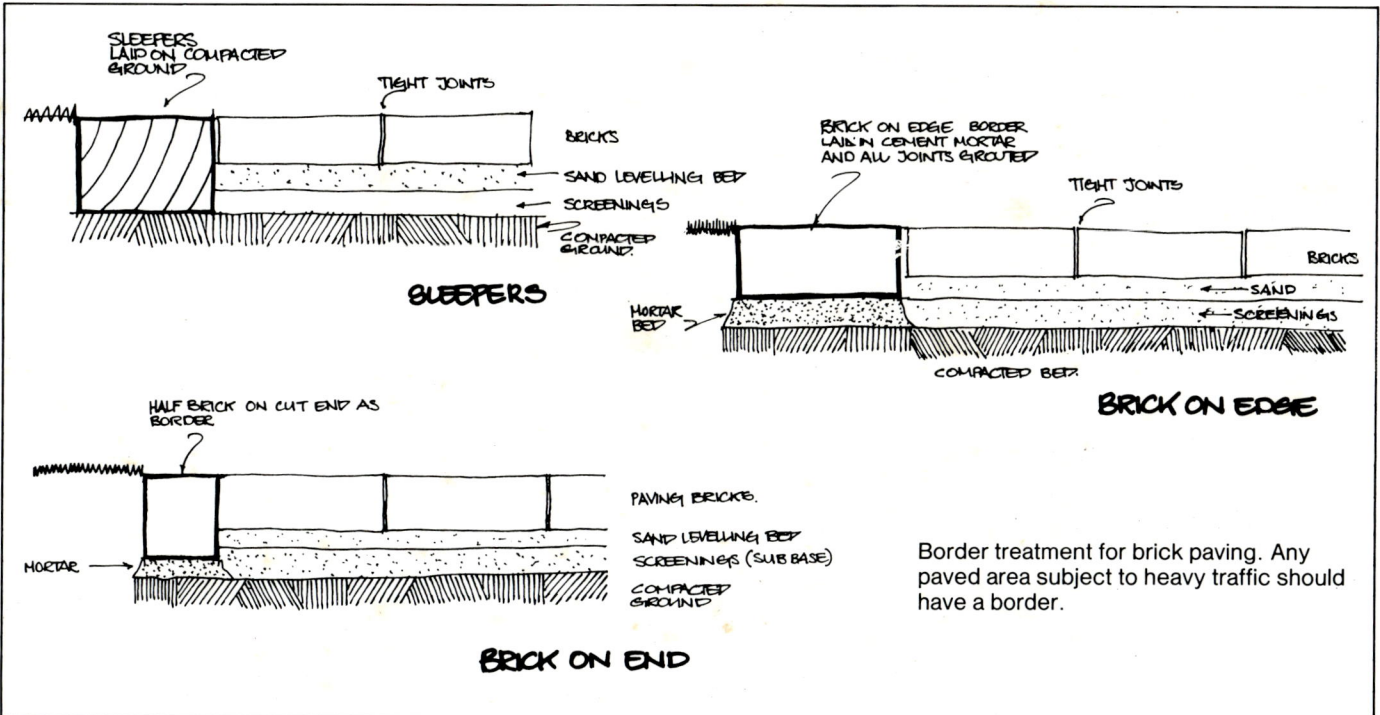

Border treatment for brick paving. Any paved area subject to heavy traffic should have a border.

Creating Grassy Verges

Where paving is to finish against lawn, the level of the paving should be at least 50 mm above the top of the grass when it has just been mowed. This will allow the paving to settle as well as providing for grass to grow. As you top-dress grass in the future, the level of the well-mowed grass will rise slowly leaving the paving too low. You need to allow for this eventuality at the design stage.

A feature border in concrete block paving.

Dry Mortar Jointing

Avoid paving styles that have open or wide gaps between the bricks. Any gap larger than 8 mm may snag the heel of a high-heeled shoe, wrecking the shoe and possibly breaking an ankle. This type of disaster only seems to happen to the most expensive pair of shoes of your most important guest.

Where you cannot avoid joints in the pattern you have chosen, fill the gaps with cement mortar. It is generally impractical to grout the joints with a wet mix of mortar and cleaning up mortar spills is tedious and often less than satisfactory. Use dry mortar joining to fill the joints. This job can only be done when the bricks and sand are perfectly dry. You will need a mix of between 4:1 and 6:1 sand and cement thoroughly mixed dry. The mix is broomed over the bricks and tamped well into all joints.

Do not attempt to sweep the sand and cement mix over the bricks while they are even slightly damp. The brick surfaces will be stained with cement which will be very difficult to remove. Wait until there have been at least several dry, warm days.

Sweep off and remove any excess mortar mix. Atmospheric moisture and any dampness in the ground will be enough to set off the mortar, but a

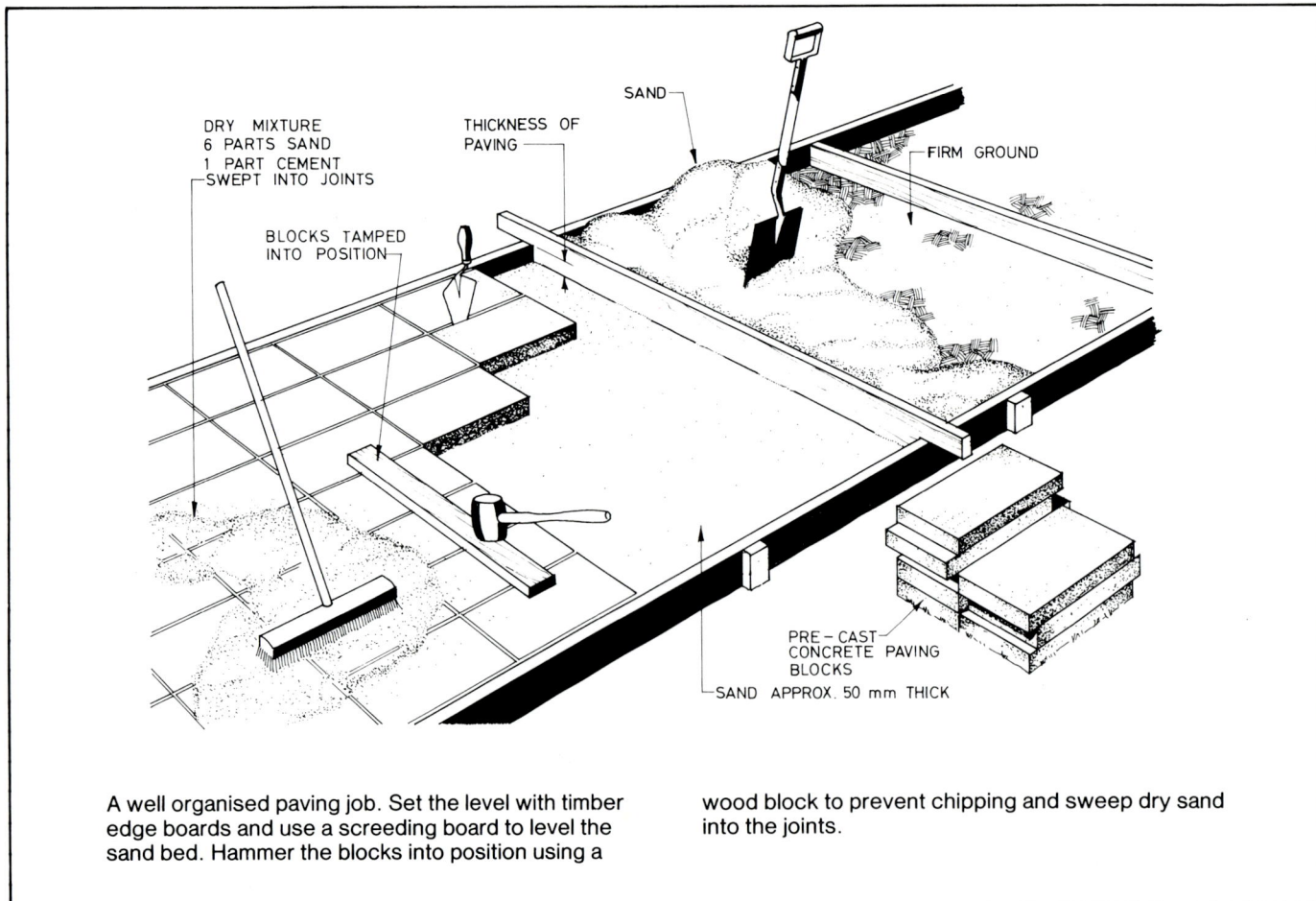

DRY MIXTURE
6 PARTS SAND
1 PART CEMENT
SWEPT INTO JOINTS

THICKNESS OF PAVING

SAND

FIRM GROUND

BLOCKS TAMPED INTO POSITION

PRE-CAST CONCRETE PAVING BLOCKS

SAND APPROX. 50 mm THICK

A well organised paving job. Set the level with timber edge boards and use a screeding board to level the sand bed. Hammer the blocks into position using a wood block to prevent chipping and sweep dry sand into the joints.

Barbecue courtesy of Barbeques Galore.

Earthenware tiles laid on the diagonal over a concrete slab.

very light sprinkle from a watering can or sprinkler is useful. Take care that no water runs across the surface of the bricks at this stage. If it looks like raining, cover the paving with plastic film.

Any mortar that has set on the surface of the bricks may be scrubbed off with muriatic acid, which you can buy from any hardware store. This strong acid is useful for removing cement stains, but can burn skin and damage your eyes. So, wear gloves as well as protecting your eyes.

Slate and Tiles

Extremely popular in recent years are slate and tiled outdoor areas which provide a luxurious and highly functional surface finish. There is a wide price range in the cost of slate tiles per square metre. The more unusual colours and thicker grades can be expensive, making the cost of laying this type of surface finish quite pricey.

Tiled surfaces are among the most attractive of hard surfaces and proper selection of materials and care in their installation will be rewarded by a lifetime of good service. Because some of the darker surfaces may become uncomfortably hot in summer months, it is important to consider shading the tiled area.

Slate tiles must be laid in a mortar bed not less than 35 mm thick and preferably 50 mm thick over a structural concrete slab. The slab will be typically 100 mm thick with light steel reinforcing mesh, similar to that required for a concrete path.

Paving works of this scale are probably better done by professional tradesmen as the labour and tools needed are not generally available to the average do-it-yourselfer. However, if you are keen don't be inhibited from attempting the job. The following instructions set out how you should approach the job. The same process is appropriate for any external tiled surface whether slate, ceramic tiles or quarry tiles etc.

51

Site Preparation

Remove all grass, root matter and topsoil so that the excavated level is about 200 mm below the desired finished level. Spread a layer of builders' sand about 50 mm thick to even out any depressions in the excavated ground and compact it thoroughly.

Position and level substantial edge boards around the perimeter of the slab. The top surface of these boards defines the final level of the concrete surface. The edge boards are kept in position by wooden stakes about 600 mm long by 50 mm to 75 mm square. These are driven well into the ground and nailed to the outside of the form boards.

If you take proper care at this stage, you will be rewarded with a good quality job. The old rule of measure twice and cut once is well worth mentioning.

While not strictly necessary, a layer of polyethylene sheet (clear plastic sheet or concrete underlay) may be placed over the sand bed and inside the forms. Plastic film underlay assists the concrete curing process and ensures that the concrete does not lose moisture to the ground during its initial setting.

Now place the reinforcing steel mesh in position. Most larger building suppliers will cut and deliver reinforcing mesh to size. Remember to allow at least 50 mm concrete cover to the steel at all sides of the slab. For example, a slab 5 m × 4 m will be reinforced by a section of mesh 4.9 m × 3.9 m after allowing 50 mm all round.

If your paved area is to serve as a driveway, you may need to ask an engineer to specify the size of the reinforcement mesh to make sure that the slab will not crack when loaded. In most small slabs (less than five metres square), path mesh of F72 mesh will suffice.

The reinforcing steel should be located not less than 30 mm from the bottom of the slab. Naturally, when the area is full of wet concrete, the exact position will be difficult to check. Professional concretors use reinforcement chairs and pans to keep the steel in the right position. Your local builders' supplier will be able to supply suitable chairs and pans and the small cost of these items will make the job far easier. About one chair per square metre will suffice.

Ordering Concrete

Once you have completed all the preparatory work, you can order the concrete. Ready mixed concrete is bought in cubic metres and costs about $120–$150 per cubic metre depending on the quantity ordered. Use local suppliers as delivery costs are related to distance travelled. Also, be sure that you have all tools, wheelbarrows and a couple of willing helpers on hand to greet the truck driver. If you are unprepared, you may be charged for any delay or your concrete may be dumped somewhere you don't want it.

Concrete is graded by strength which is measured in megapascals. The shorthand for this rather intimidating term is MPa. All you need to know is about 20 MPa concrete is adequate for all domestic work.

In theory, for a 100 mm thick slab, one cubic metre of concrete will cover an area of ten square metres. In practice, the rule of thumb is that one cubic metre will give an effective coverage of about eight square metres.

Although 20 per cent wastage may seem excessive, one barrow load short will cause problems. It's better to waste a little than be caught short. So, assuming that the slab is to be 100 mm thick, you should order one cubic metre per eight square metres of slab area.

While you are waiting for the concrete truck to arrive, make sure you have the following at hand:

- ☐ 2 barrows, one to load from the truck and the other at the job site
- ☐ 2 shovels
- ☐ 1 rake
- ☐ 1 screeding board as long as the paving is wide
- ☐ 1 pair gumboots for concretor (you)
- ☐ 1200 mm level
- ☐ Trowel, edging tool and jointer
- ☐ A couple of willing assistants
- ☐ A hose and a 20 litre drum of water
- ☐ Some cement (1 bag will do)
- ☐ A clear access from delivery point to the job

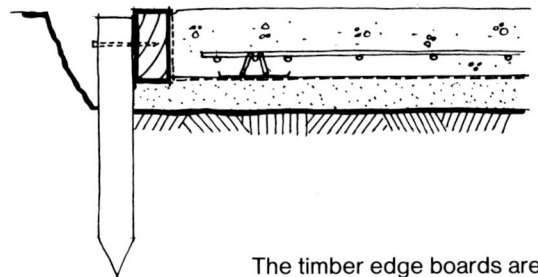

The timber edge boards are fixed in position with stakes driven firmly into the ground.

Placing and Compacting

Once the truck arrives, place the first lot of concrete at the spot furthest from the delivery. Remember, you are going to get tired and you can't push your wheelbarrow across poured concrete.

Make sure the reinforcement is not moved while you place the concrete and take particular care to compact the mix properly. Tamp the wet concrete down to remove any air bubbles. On large jobs, professionals use an electric vibrator, but on a job of this scale, using the head of a rake in a vertical motion should be sufficient.

Once you have placed the concrete, level it out with a screeding board. This long piece of timber is

A highly detailed area using slate and terracotta tiles.

used with a crosswise motion and dragged towards the concretor to fill in voids and scrape off high spots. Vertical tamping forces the aggregate below the surface and helps proper compaction.

If you are going to lay tiles or slates in a cement bed, then the rough surface is good enough to ensure adequate adhesion. Use the edging and jointer tools to tidy up the edges against the timber forms to assist in their removal.

Remember that overworking the surface of the concrete may result in poor bond and dusting later. Once your concrete slab has achieved its initial set, keep it moist for not less than seven days. Cover the slab with plastic sheet, sand or simply hose it down regularly. Wetting during the curing process is particularly important for any slab that will be a base for tiles as surface quality is critical for adhesion.

Leave slab for at least two weeks before laying the tiles, longer if possible, to make sure curing is complete and initial shrinkage has ceased. Before tiling, soak the slab, but surface water should not be present while laying tiles. And remember, make sure to follow precisely the suppliers' instructions when laying the tiles. As a rule, slate, ceramic and quarry tiles should be laid in a mortar bed not less than 35 mm thick.

Mortar for Tiling

A suitable mortar mix for tiling of all types may be mixed from clean plasterers' sand and Portland cement in the proportions of 3:1 by volume.

A 25 square metre paved area requires almost one full cubic metre of mortar. Once again, the use of a mechanical cement mixer is strongly recommended. The mixer will assist in consistency of mix quality and colour, and speeds up the work.

Laying Tiles

The use of a good quality builders' level is essentially for all tile laying. The longer the level the better — 1200 mm is suggested.

Mix only sufficient mortar for the amount of tiles that you can comfortably lay at one time. Use a bricklayer's trowel to place the mortar and tap each tile into position carefully ensuring that the mortar bed gives full support to the whole tile.

Take your time and constantly check levels and falls. Don't be afraid to remove a tile and re-lay it if the levels are not correct. The secret of success is care and slow, methodical work. Specialist publications and tile suppliers will be able to give you additional information.

5

PERGOLAS

Footings • Structure • Connections • Bracing • Coverings • To Paint or Stain?

One popular method of linking your outdoor barbecue area with the remainder of the house and garden is with a pergola. In its simplest form, a pergola is a timber frame erected over the ground to shade and shelter.

By combining good design with the right vegetation, you can build a very effective semi-enclosure. Most people feel more comfortable when eating outdoors if protected in some way. A pergola gives this feeling of protection without all the restrictions associated with a full roof covering.

Pergolas can also bring considerable benefits to your house by shading and protecting problem windows. Some designers prefer to cover pergolas with shadecloth, while others let nature do the job with deciduous climbers such as ornamental grape, bougainvillea or wisteria. These plants grow luxuriantly during the spring months and lose their leaves in winter, allowing the weaker winter sun through.

Pergola design is quite simple and the techniques of construction are well within the capability of the enthusiastic do-it-yourselfer. The detailed design depends on the aspect of the sun in relation to the area you want to cover. In the southern hemisphere, most people choose to have the northern walls facing the outdoor living area. In this case, the midday and afternoon sun in summer makes conditions too hot for comfort, especially at the times when most meals are eaten outdoors.

An effective pergola for this situation will have the basic frame members running north to south and the shading members closely spaced, running parallel to the north wall of the house. Western facing pergolas require the main structure to run east to west with the shading battens placed in the north to south direction. Shadecloths can supplement the shade of the timber work.

There are five basic elements in designing a pergola: footings, structure, connections, bracing and covering.

Footings

There are several types of commercially available shoes you can use to anchor the posts of the pergola. Naturally each pergola requirement will be determined by individual circumstances. However, the following suggestions will provide you with a guide.

Depending on the load that the footing will be required to hold, a concrete pad 300 mm in diameter and deep enough to penetrate to undisturbed ground will suffice in most circumstances.

If the footing is to be put into sand and it's difficult to stop the sand from collapsing into the hole, a clever trick is to cut the top and bottom from a 20 litre drum and use it as the formwork for the pad.

There are a variety of methods for connecting the posts to the footing. The object is twofold: to form a solid and simple structural connection of the post to the footing; and to avoid posts rotting in the ground. Most hardware outlets and builders' suppliers stock a complete range of connectors that will suit most circumstances.

You can make a simple homemade connector from a length of 75 mm × 12 mm galvanised arch bar cast into the pad and carefully housed into the post. This arrangement suits almost all situations and has the advantage of accommodating all types of pavings. (See diagrams.)

Your local hardware supplier or builders' yard should be able to supply these simple connectors. If you can't obtain them, buy the arch bars and ask a local metalworking shop to drill the holes and cut the steel to size. To prevent rusting, make sure all metal shoes have been hot dip galvanised and all bolts either cadmium plated or galvanised.

Similarly, make sure all nails you use to build a pergola are galvanised and well punched into the surface. Then plug the hole with putty or wood filler. Rusting nails eventually stain the timber and could fail completely. The extra cost of using galvanised nails is inconsequential when compared to the cost of replacing the structure.

Most commercially available connectors are designed to be bolted onto concrete slabs and are not ideally suited to unit paving. An ugly concrete pad at the surface level of the paving may be required. If you intend installing paving blocks, then your choice of footing and connector should allow for the extra depth required.

If you use the arch bar connector method, cut your connectors from 900 mm lengths to give two 450 mm pieces. Drill two 10 mm diameter holes, 100 mm apart at the centres, 50 mm and 125 mm from the top edge. The top of the connector should be the galvanised end rather than the cut end.

If you can place the cut end into the concrete pad, then you will avoid rusting. You can use this type of connector to fix pergola posts to the edge of planters and most other situations, and it should be cheaper than most commercial fixings.

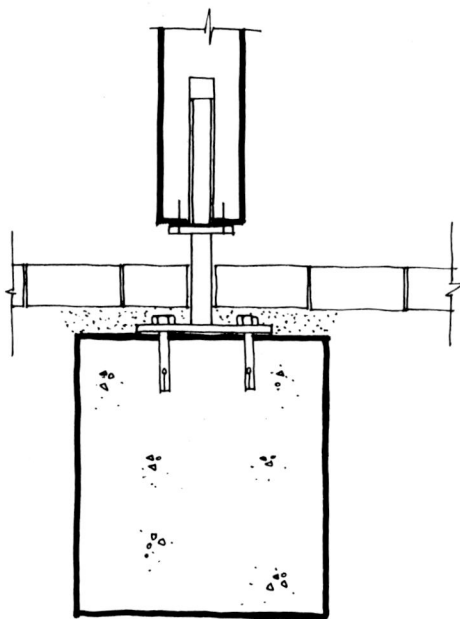

POST CONNECTOR TO MASS CONCRETE FOOTING

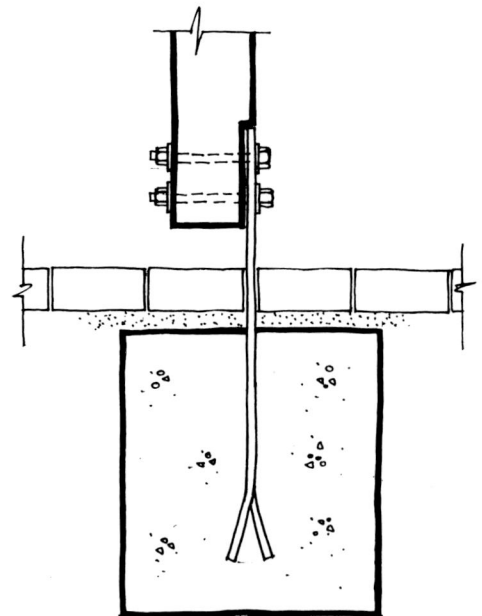

POST CONNECTOR TO MASS CONCRETE FOOTING.

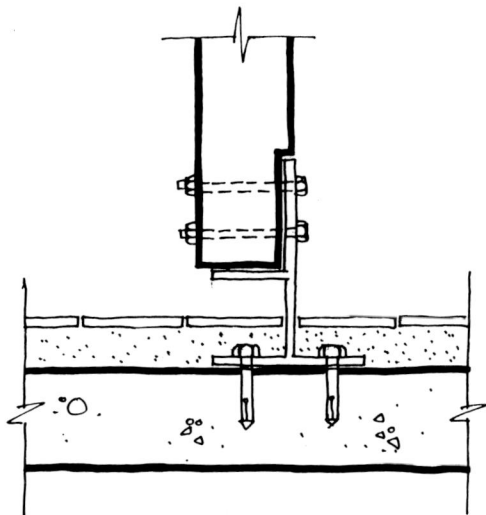

POST CONNECTOR TO
CONCRETE SLAB.

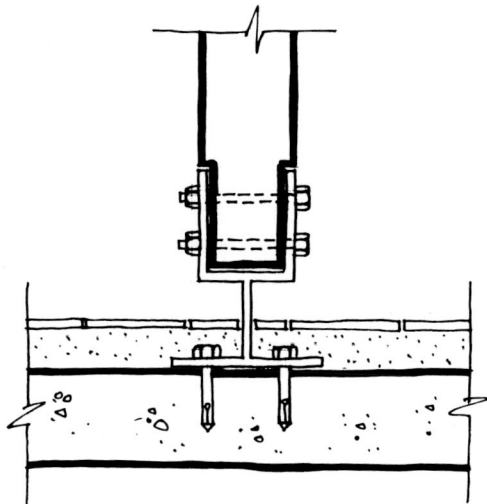

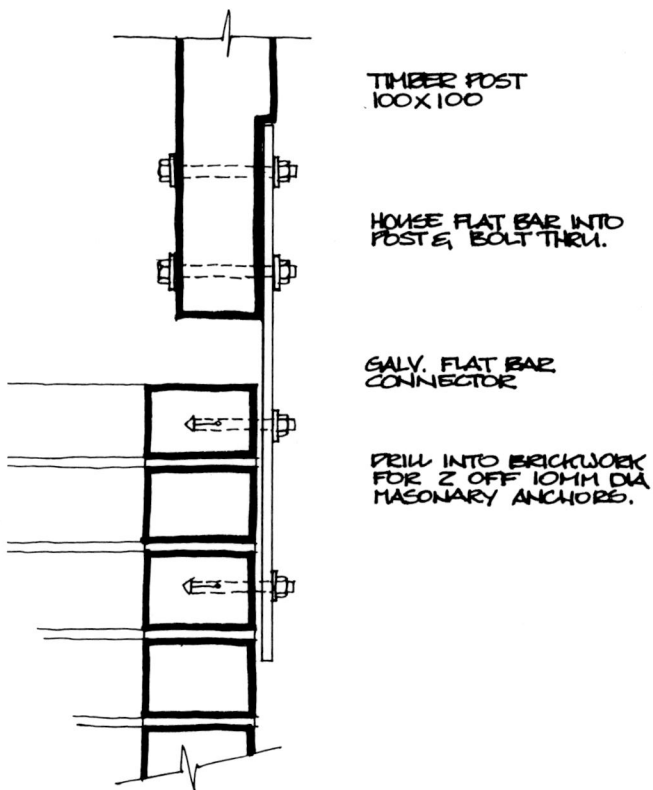

POST CONNECTOR TO
CONCRETE SLAB

TIMBER POST
100 x 100

HOUSE FLAT BAR INTO
POST & BOLT THRU.

GALV. FLAT BAR
CONNECTOR

DRILL INTO BRICKWORK
FOR 2 OFF 10MM DIA
MASONARY ANCHORS.

100x100
TIMBER POST

GALV. METAL
TEE CONNECTOR
BOLTED THRU

PAVING

CONCRETE
FOOTING

POST CONNECTOR TO
BRICK DWARF WALL

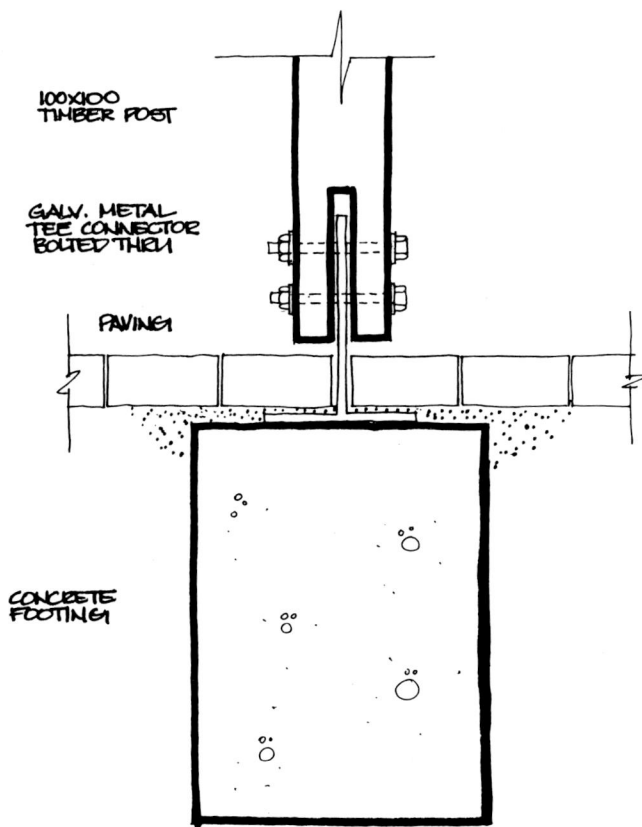

POST CONNECTOR TO
CONCRETE PAD FOOTING

Structure

Square timber posts for pergolas may be either from 100 mm × 100 mm timber or, for smaller pergolas, 75 mm × 75 mm. Straight-grained, well-seasoned timber is particularly important, as any warping and twisting is especially noticeable on a pergola.

The choice of timber is largely up to you. Oregon or Australian hardwoods are both suitable although Oregon is more easily worked. However, it is more likely to develop cracks as it dries out which is why the rough sawn grades of timber are more suitable for pergola construction than dressed (planed smooth) finished timber.

The foot of each post is housed in the post connectors and the bolt size should be selected to minimise the length of thread exposed. This will help reduce rusting and save many a grazed ankle bone. The technique of counterboring the hole to a depth sufficient to conceal the nut and washer is a good design practice.

Left: A well designed pergola gives a feeling of enclosure and provides shade.

Right: Deciduous climbing plants provide shade in summer but allow the sun to penetrate when they lose their leaves in winter.

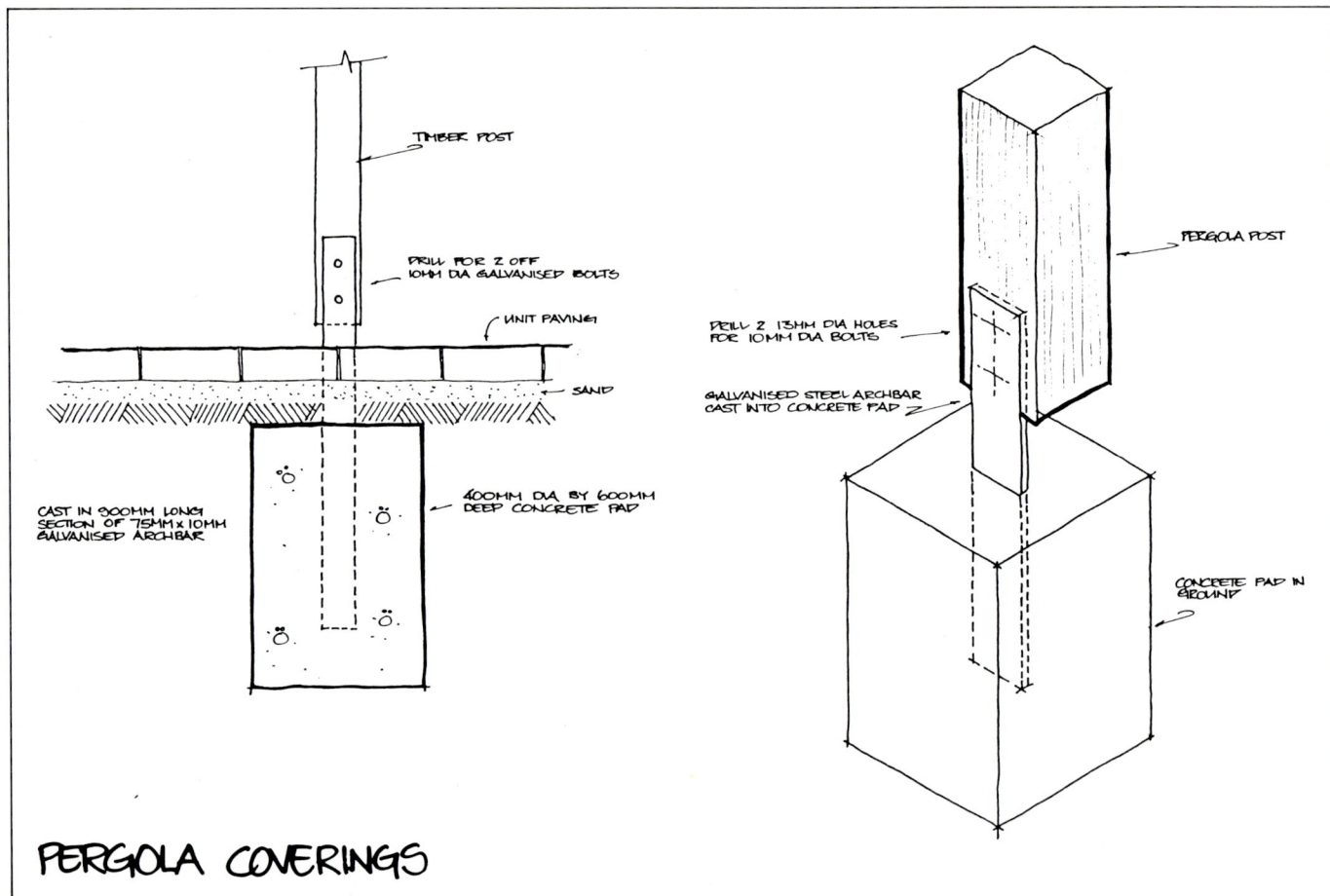

TIMBER POST

DRILL FOR 2 OFF 10MM DIA GALVANISED BOLTS

UNIT PAVING

SAND

CAST IN 900MM LONG SECTION OF 75MM x 10MM GALVANISED ARCHBAR

400MM DIA BY 600MM DEEP CONCRETE PAD

PERGOLA POST

DRILL 2 13MM DIA HOLES FOR 10MM DIA BOLTS

GALVANISED STEEL ARCHBAR CAST INTO CONCRETE PAD

CONCRETE PAD IN GROUND

PERGOLA COVERINGS

Pergolas are great for hanging plants.

Connections

In carpentry, a horizontal load-bearing member is called a plate. In pergola design, a plate is required at the house wall and another supported by the line of posts.

If your house is framed up in timber studwork (weather-board construction) then the wall plate may be nailed or screwed through the wall cladding directly into the timber frame. To locate the studs behind the cladding, first look for the vertical lines of nails.

If you cannot find these nails, you may be able to locate the studs by tapping the wall and listening for the solid sound. A device called a studfinder is available from your hardware supplier to help with this task.

Before you drill or nail into a wall, make sure you are not going to hit a water pipe, gas line or electrical cables. In masonry walls, the connection of the wall plates may be done using wall anchors. There are several types available and all work on the expansion principle.

Simply drill the required diameter hole into the wall to a sufficient depth to take the anchor and insert the device and tighten with a spanner. To conceal fixings, the experts locate them in the housings for the rafters (see diagram) or counterbore the holes in the timber to leave the bolt heads flush with the surface.

The plate or beam at the outside of the pergola is supported by posts. The correct way to fit this beam is to house it into the top of the post. This type of joint can be cut on the ground before the posts are erected and gives temporary support while the bolt holes are drilled. If you are working single-handed, a couple of G clamps will make the job safer.

A housing joint is one where one member is cut out, in order to allow the cross member to fit into the recess. To mark out this type of joint, simply place and clamp the cross member in position and mark on the post where the edges cross. Mark the depth as required (generally not more than half the thickness of the thinner member, the plate) and cut carefully with a tenon saw, **inside the lines**.

CONNECTING PLATES TO BRICKWORK

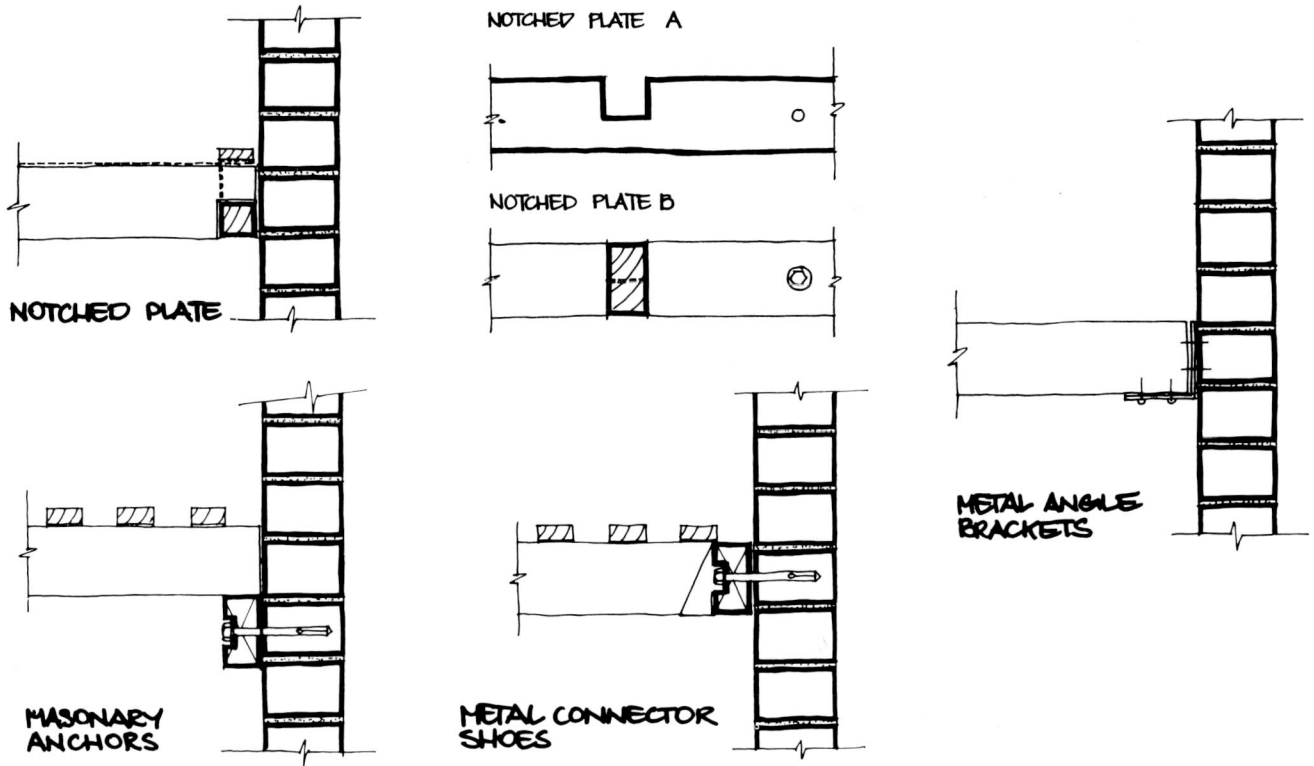

NOTCHED PLATE A

NOTCHED PLATE B

NOTCHED PLATE

METAL ANGLE BRACKETS

MASONARY ANCHORS

METAL CONNECTOR SHOES

PERGOLA DETAILS

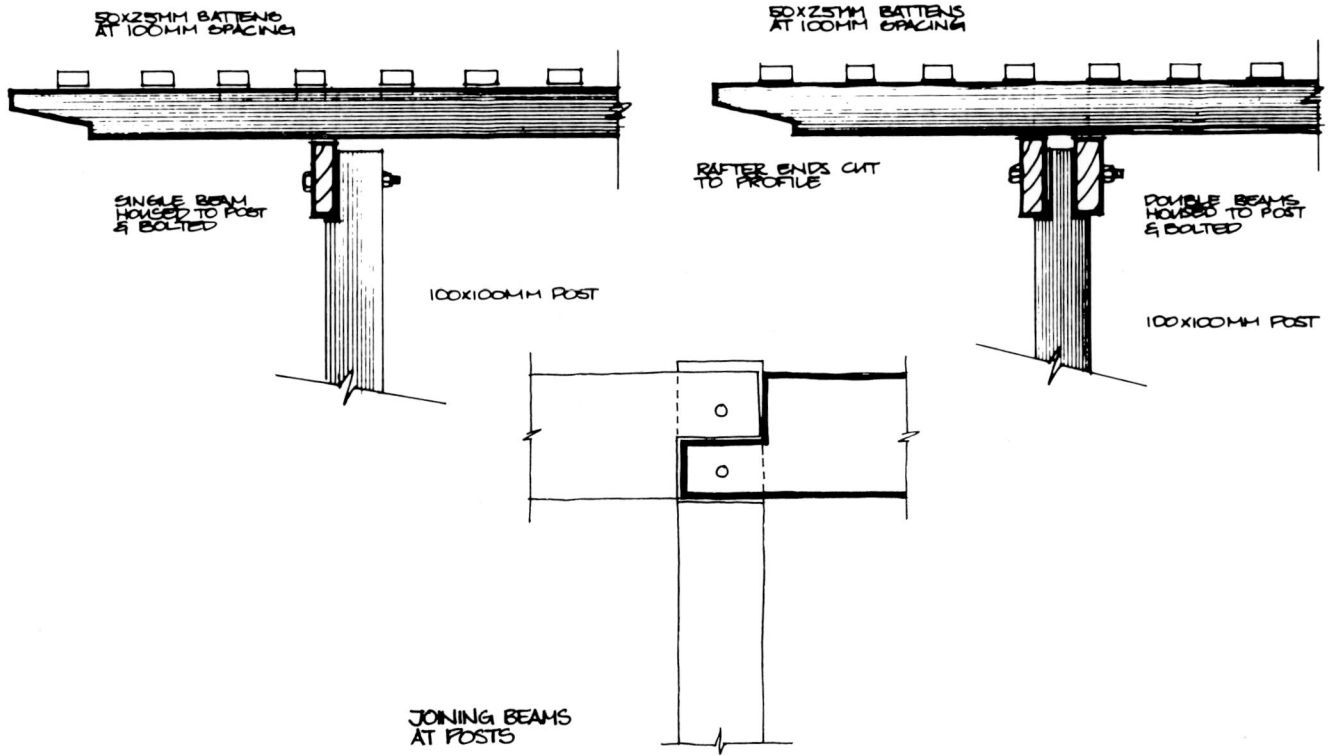

50x25MM BATTENS AT 100MM SPACING

50x25MM BATTENS AT 100MM SPACING

RAFTER ENDS CUT TO PROFILE

SINGLE BEAM HOUSED TO POST & BOLTED

DOUBLE BEAMS HOUSED TO POST & BOLTED

100x100MM POST

100x100MM POST

JOINING BEAMS AT POSTS

Remove the excess material with a chisel and make the joint smooth with a paring action. The result will be a professional-looking joint that will form a strong and effective connection. As a rule, two 10 mm diameter galvanised steel bolts recessed in counterbored holes will complete the connection.

It is a good idea to drill the holes when the beam and posts are in their final position with all posts plumb (vertical) and the plates set dead horizontal. Again, keep in mind the old carpenter's rule of measure twice and cut once.

Bracing

Unless the pergola is built inside two walls at right angles to each other, you will need some form of bracing to stabilise the structure. This may take the form of knee braces between the posts and rails or diagonal braces in the plane of the rafters.

For design reasons, carefully consider the use of diagonal braces. An effective and good-looking diagonal corner brace may be inserted at an angle of 45 degrees, fixed between the double beams and attached to the posts. The brace will need to be carefully dimensioned so as to fit neatly between the beams, secured with galvanised nails, or preferably galvanised steel bolts, and nailed at the post (see diagram).

When setting out for this brace, the double beams must be housed into the posts sufficiently, so that the space between them will be equal to the thickness of the angled brace.

A less satisfactory method to avoid braces is to cast the posts into substantial concrete footings and rely on the cantilever effect to stabilise the posts. This method is best suited to free-standing pergolas and only if treated timber has been used.

Coverings

The covering of your pergola has to be considered in relation to its location and aspect.

A north-facing pergola should be designed to filter the higher midday sun whereas a west-facing, outdoor living space will need to control the rays of the lower, afternoon sun. If the pergola is properly designed, it will form part of the whole design of the outdoor living spaces and control the sunlight in a way that will make the area the most comfortable environment of the home.

You can choose a combination of rafters and battens, sometimes with shadecloth coverings, to control the sun. Depending upon the aspect of the structure, you can space the rafters or battens close enough to block out the sun at the required time of day.

A simple design is a north-facing pergola with a

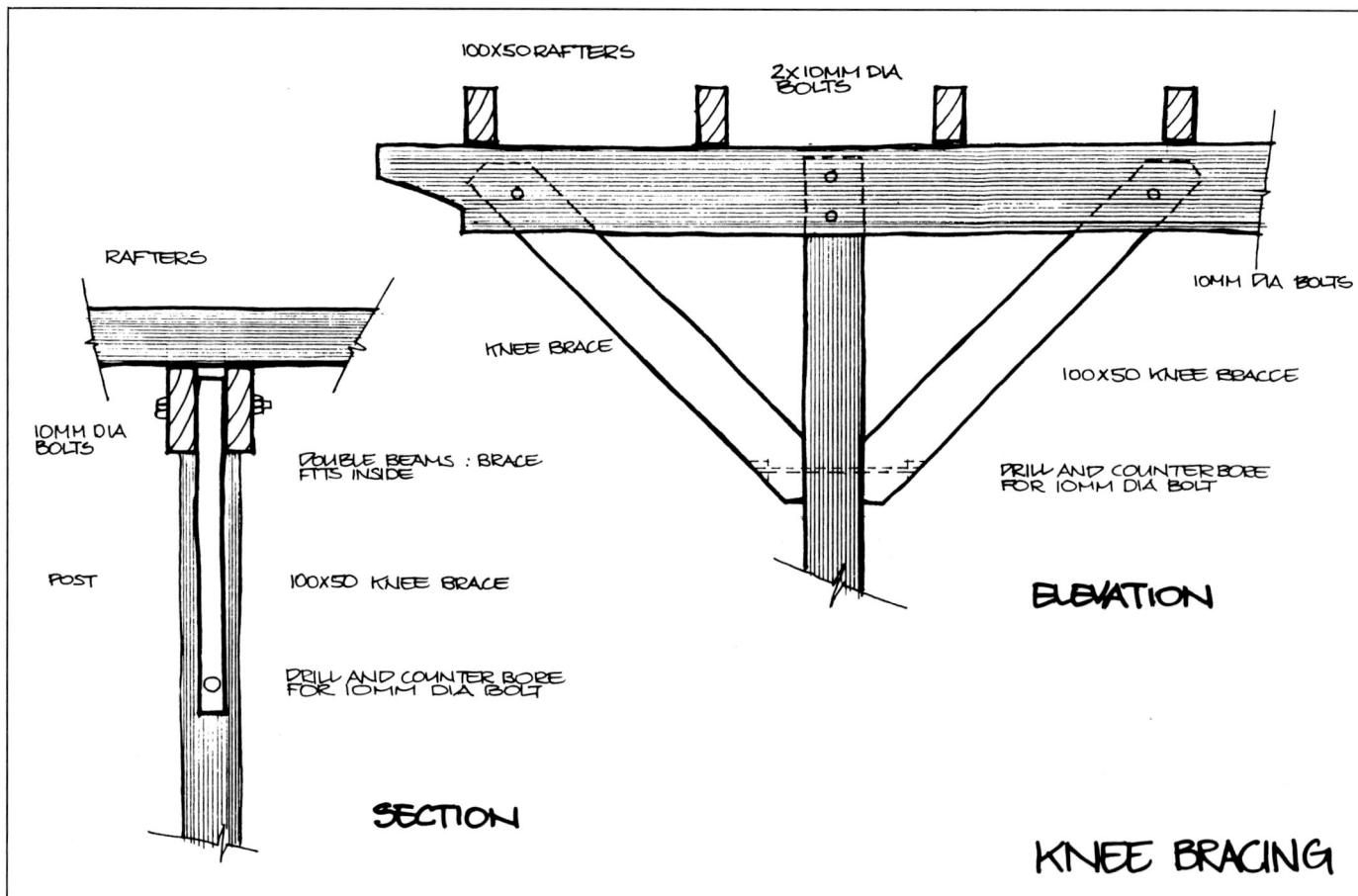

KNEE BRACING

line of posts parallel to the house, supporting a pair of horizontal beams. These, in turn, support a line of rafters, spaced at 600 mm. The minimum height of a pergola should be no less than 2400 mm, and to be effective will be 3 m to 4.5 m deep.

At midday when the sun is due north, sunlight will penetrate. By 2 pm the sun will be almost completely blocked and the area shaded. Remember that the pergola will have to be considerably larger than the size of the outdoor living area to provide complete shade and shelter. So, you will be better off to allocate an area for your furniture and design accordingly.

The horizontal spacing for rafters depends on their height. If you use 100 mm × 38 mm, spacing the rafters should be reduced to about 400 mm apart. For rafters 150 mm deep, the spacing should be about 600 mm. As in all design matters, you have to make compromises. A pergola, unless it has movable parts, will sometimes let some sunlight through.

It is most important to ensure that when some glare comes through, it is not going to cause problems. The difficulty here is that afternoon sun in late autumn and winter is most welcome while in mid summer, it is a nuisance. Therefore, you must take

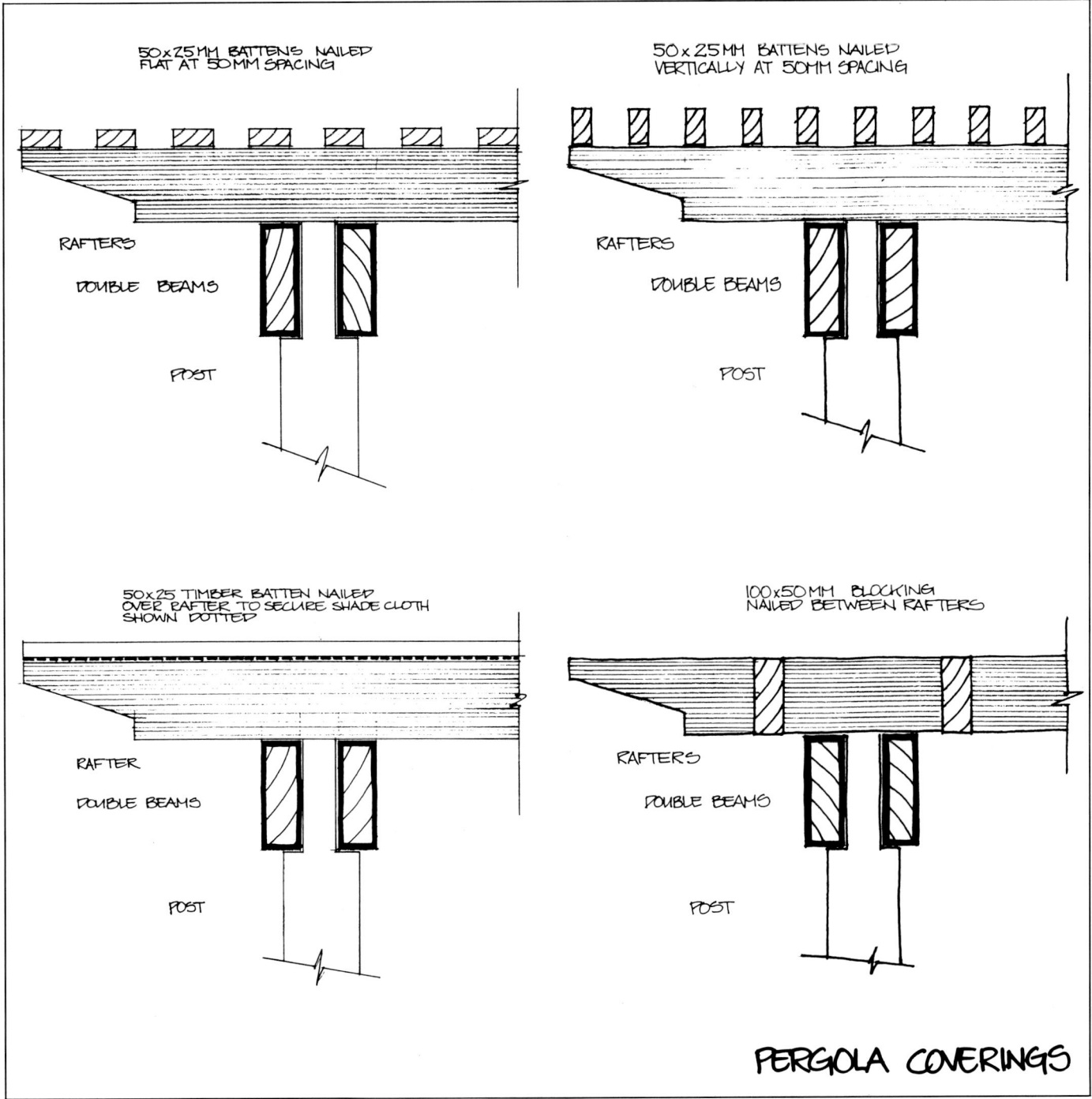

PERGOLA COVERINGS

This painted pergola gives a light and airy feel to the garden.

into account the times the outdoor areas will be most often used and design accordingly.

A way of saving on the cost of expensive timber is to substitute small battens, say from timber 50 mm × 25 mm, laid over rafters spaced as far apart as 900 mm. You should also remember that there is a distinct possibility that kids will climb your pergola. Therefore, all sections must be strong enough to support a person's weight. A delightful and interesting pergola covering can be created with small diagonal battens. This method if done carefully, provides a highly effective sunscreen which is also visually exciting.

Shadecloths are a comparatively recent innovation for pergolas. These woven plastic fabrics provide excellent service and come in a variety of densities, according to the percentage of light transmitted. The most useful for the pergola builder will in be the range of 80 to 90 per cent.

You can secure these fabrics to the tops of the rafters with staples or timber battens. In any case, all nails, screws and staples must be heavily protected against rusting, preferably by galvanising.

Pre-treated timber lattice work is becoming increasingly popular for pergolas for both screening and covering. The advantage of pre-treated timber is that it will resist the infestation of termites and borers and should last the life of the house. Some brands of timber on the market come with a 20 year guarantee.

It may be worth considering closing the western end of a pergola to reduce the glare and heat of the afternoon summer sum. This can be done easily with a trellis, lattice work or by planting appropriate shrubs.

Remember, a barbecue and outdoor eating area is most likely to be used in the afternoon and evenings, so the design of the shading of your pergola is crucial to the total enjoyment of your outdoor living and entertaining.

To Paint or Stain?

To protect the timber from discolouration and decay your pergola requires a protective finish of paint or stain. It is important to remember that once the pergola has been built and climbing plants are established, painting will be a difficult job. The most sensible time to paint a pergola is when the timber is on the ground **before you start building**.

When you paint or stain timber on the ground you're able to give a very generous coat of stain without the problems of painting overhead. You won't need any drop sheets and the job will be done in a fraction of the time.

There are two methods to consider. First you can stain all the timber as soon as it arrives, build the pergola and touch up where joints have been cut into the timber. The second is to cut each part to size, make all notches, joints and end finishes and then stain each completed part before you build. The method you choose depends on how much of a hurry you're in, once you start the cutting of the timber.

For external timber work you have the choice of two basic finishes:
☐ Conventional paint finishes
☐ Stain finishes

Whether you use paints or stains depends on your personal preference and the style you have created in your living area. If your taste is informal and the surrounding buildings feature lots of exposed timber, then a pergola made from rough sawn timber, finished with a natural brown stain would be appropriate.

If your house has a contemporary design, then your pergola may be better constructed from dressed (smooth) timber and finished with one of the colourful, water-based acrylic paints. A huge range of colours is available to match almost any style.

Coloured finishes may be a little longer lasting than the stain types, but refinishing can be difficult. Consider all aspects of design, final colour and finish before you choose what suits you and your home best. The old tradesman's rule of do it once, do it right applies especially when you are building and painting your pergola.

Right: Delicate plants can flourish under the protection of shadecloth fixed over a pergola.

6
DECKS

*Council Approval • Design •
Construction • Footings • Timber Posts •
Bearers and Joists • Connecting the Deck
to the House • Timber Decking • Railings
and Stairs*

A timber deck is an excellent substitute for paving especially if your house is built on a sloping site. If the living areas of the house are separated from the ground level by steps, the cook will soon tire of climbing up and down for items needed at the barbecue area.

A well-designed and integrated timber deck solves the problem of a sloping site and steps, and provides valuable outdoor living and entertaining space all year round.

Incorporating the barbecue into the design of a deck or terrace area is no less important than the design of the ground level paved area. It is unlikely that you will be able to build a heavy, brick barbecue at the high deck level and a portable barbecue unit may suit you better. However, if your barbecue can be built into the walls of the deck area without much trouble and without too much expense, then the results will be very effective.

Council Approval

The techniques of building a timber deck are only a little different from a pergola. But there's one important difference. An attached timber deck is regarded as a structure by most local councils and they will require you to submit a building application for approval.

Anyone contemplating building a deck attached to the house should find out from the building inspectors at the local council whether a building application will be necessary.

Your council officers will tell you of the procedure to follow but more than likely you will have to prepare a plan and specification showing the location and details of the proposed deck. The sizes of the timber structural members must comply with Australian Standard 1684 SAA Timber Framing Code published by the Standards Association of Australia. This can be referred to at the Association.

The sizes given here are a guide only and should be checked for each individual design. The Timber Development Association in your State will be able to give you advice if you are contemplating building a timber deck. Their advice can be particularly helpful in selecting timber.

Remember local councils can order the demolition of unauthorised structures so for this reason alone, follow the proper procedures. Your drawing submitted to council need not be a great work of art; a simple, clear sketch plan and elevation will generally be enough, showing any special design features.

Design

When you construct a timber deck you are building something that is broadly similar to the floor system of the house itself. However, because the deck is constantly exposed to rain and sun, the choice of timber is of great importance, as it must resist warping and rotting.

A deck must also be designed to take advantage of all natural benefits such as sunshine and breeze. This means a careful analysis of site conditions.

As a general rule in the southern hemisphere, a good location for a deck is on the north side of the house where the longest hours of sunshine are received. It may also be necessary to provide some protection with a pergola or verandah extension, from the hottest summer sun.

If you are fortunate enough to be able to build your deck so that it turns the north-western corner of the house you'll find there are several advantages including the ability to move to the more shaded northern section when the western sun gets too hot.

A large or very high timber deck may be better left to a professional builder as the problems of height plus special equipment such as stages and scaffolding may make the job uneconomic for the do-it-yourselfer. Building a reasonably sized deck, however, is well within the abilities of a well-equipped handyman.

For a more involved deck, it may be useful to ask an architect or professional draughtsman to prepare the drawings and specification for submission to your local council. Always get a quotation from your designers before they start, to avoid arguments later. Remember, with most professionals what you pay for is what you will get.

The *Yellow Pages* of the telephone directory lists architects and architectural draughtsmen. Select one close by. The Royal Australian Institute of Architects has branches in all capital cities and they will give you advice on choosing an architect.

Construction

The floor structure of your deck will be designed similarly to an internal timber floor inside your house. The deck must be braced against the house as well as diagonally to prevent swaying.

Depending on the type of subsoil, any timber posts will require substantial footings cast from concrete. The size of the footings will range from a simple 300 mm × 300 mm pad, 100 mm thick to level out a rock foundation, to maybe 600 mm × 600 mm by 400 mm deep for a larger deck on a poor foundation such as sand or clay.

Because each site is different, the advice of an experienced structural engineer may be required. Look around your local neighbourhood and see what other homeowners have done to solve their design problems. In this way you'll be able to submit a design to council that will be approved first go.

The structural framework of the deck consists of footings, timber posts, bearers and joists and the timber decking. You also need to consider a balustrade, railings and in most instances, a stair.

Footings

The first step is to mark out the positions of the footings that will carry the load to the ground. Dig holes for each footing pad sufficiently deep to reach undisturbed ground. No roots or other vegetable matter should be present and the base of the hole should be well tamped down.

If the soil is self-supporting, you can pour the concrete straight in, but it is better to build a timber formwork to retain the wet concrete and provide a level edge to finish to. When you build the formwork,

Right: A skilful design has maximised the advantage of this sloping site.
Barbecue courtesy of Barbeques Galore.

use waterproof plywood rather than chipboard and remember to leave the nailheads clear of the surface of the timber — this will help their removal when demounting the form. Where the site allows, try to set the tops of all pads to the same level by using a string line and long spirit level.

Build the formwork at the workbench and lower it into position, taking care not to disturb the bottom of the hole. In Australia, it is preferred to set timber posts above the finished level of the concrete pads to avoid rotting of the post bases and termite attack. There are a wide variety of commercial post bases, designed for decks and pergolas at any building materials supplier. (See also Pergolas.)

Galvanised steel post connectors may be cast into the top of the concrete pad or fixed with drilled holes and masonry anchors. In general, the cast-in type will give a neater result, but the initial positioning may complicate pouring and finishing of the concrete. For neatness, house the base of the post in the support and counterbore any holes to conceal boltheads. Always use heavily galvanised bolts etc. for external structural connections. The consequences of a bolt rusting through could be catastrophic!

Timber Posts

For the posts of your deck you can use either rough sawn Oregon or dressed Australian hardwood. Both types take stain well and will match the timber you use on the deck. Make sure to use only straight-grained, well-seasoned timber for the posts.

Insist on selecting the best pieces from the stack at the timber yard and be prepared to reject any timber with splits, large knots and peculiar grain. You may have to pay a premium for asking for select grade rather than merchantable but the better quality timber is worth a small extra cost.

Hardwood posts are somewhat more durable than Oregon but this advantage is offset by the easy workability of the softer timber. Some green hardwoods tend to bleed and can stain concrete paths and paving.

A technique called stop chamfering along the exposed edges of all timbers used in the deck will help control splitting and splintering of the edges of the timber. Be sure to plane with the grain as you pass the hand plane, for a couple of strokes, along each edge of the timber. Stop chamfering gives a professional finish to your work.

Bearers and Joists

Once the posts have been erected, they will have to be propped into the correct position for marking out the joints for the bearers and joists. You can achieve a strong and effective structural connection by half-housing the bearers into the posts (see diagram).

The best way for the do-it-yourselfer to achieve a professional result is to erect the posts temporarily, to assist marking out the joints accurately. By careful use of the string line and spirit level, you can establish the level of the top of the bearer at each post.

The posts can be removed one by one and the joints cut. Because in most cases, the posts also support the handrails, leave the ends until the flooring is installed for finishing.

The bearers (usually from 150 mm × 50 mm timber) are clamped into position, accurately set up so that the posts are all vertical, the holes drilled and the galvanised bolts inserted and tightened. At this stage, the structural frame is complete and it is a good idea to touch up any areas with stain that may be difficult to paint later.

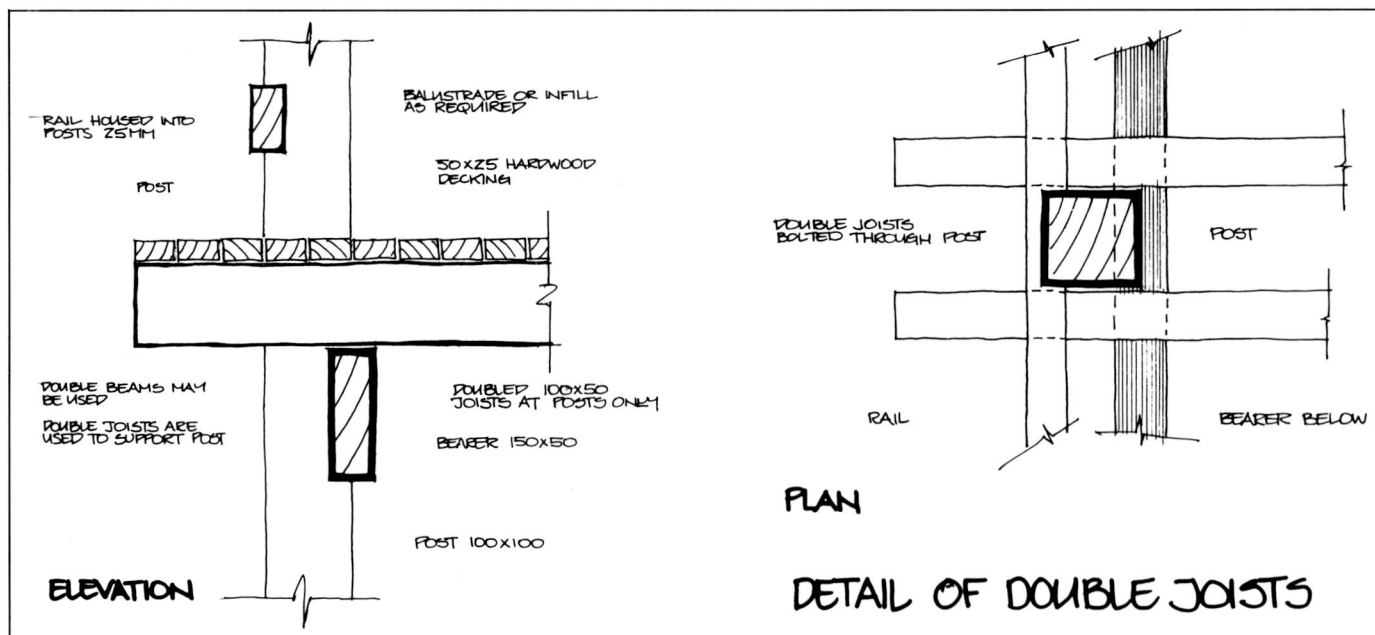

ELEVATION

DETAIL OF DOUBLE JOISTS

PLAN

Connecting the Deck to the House

It is well to remember that a large deck will carry a significant load when you are entertaining so the structure must be designed to allow for the maximum number of people likely to be standing on it. One person per square metre is a useful guide. An 8 m × 5 m deck with 40 people 'on board' could have a load of three tonnes.

The member that takes the load at the house wall is called a wall plate. Of similar sized timber to the bearers, the wall plate may be attached to the house structure by coach bolts if the house is timber, or by masonry anchors if the house is brick. Make certain that the plate is properly fixed to the walls of the house, taking care that the connections have been carried in accordance with the manufacturer's instructions. It is vital that the work is done correctly because the deck has to take a substantial load.

Use the longest practical bolts or masonry anchors; penetrating the structure to a minimum of 100 mm. Therefore, if your plate is 50 mm thick, you should use a bolt or anchor of not less than 150 mm. Set out the plate so that the finished floor level of the completed deck is no closer than 50 mm below the internal floor level. While flush floors may be desirable, the joint is difficult to waterproof. A step can be useful.

The size of the floor joists will depend on the span you require. The SAA Timber Framing Code defines the sizes needed, but as a guide, a span of less than 2000 mm and a spacing of 450 mm will allow the use of 100 mm × 50 mm Oregon joists. It is often preferable to use a doubled joist at each post and at each intermediate baluster. The doubled joists in these locations will assist the installation of the balustrade railing and greatly increase its strength.

Timber Decking

Although the choice of timber decking depends to a large extent on personal preference, the use of well seasoned hardwood is recommended to resist wear and tear. Your design should allow the timber some freedom to move as it gets wet and dries out again. A small space between the boards is all you need.

Keep this gap to a minimum to avoid catching the high heeled shoes. The ideal gap is between 2 mm and 3 mm. When you are nailing the decking down, it's useful to have a thin strip of timber planed down to the required thickness as a guide. Simply slip the strip of timber against the previously nailed-in piece, then hold the next one tight against it while you drive the nail in.

Hardwood decking can be bought from the larger timber yards in 50 mm and 75 mm widths and a thickness of 25 mm. These sizes are prior to finishing, a job which usually removes about 5 mm from each dimension. When you calculate the amount of timber required, ask your supplier what the finished size of the decking will be.

Only hammer in one nail at each joist. Two nails will cause splitting as the timber shrinks. Drill for nail fixings at the end of boards to prevent splitting. The nails should be galvanised and punched well below the surface. Fill all holes with wood putty.

You can install decking across the joists conventionally, or on the diagonal for a better effect. If you decide on the latter you should reduce the spacing of the joists as the decking will be spanning over a great distance which will increase the deflection. While it is essential to leave spaces between the floorboards to prevent water from ponding, it is also necessary to treat the ground below in some way to ensure that the run-off will not go under the house and undermine the foundations or cause dampness problems. Some people prefer to use pebbles over plastic sheeting as a ground cover under the deck because this is an area where restricted access will make maintenance difficult. Whatever your solution, you must make provision for drainage.

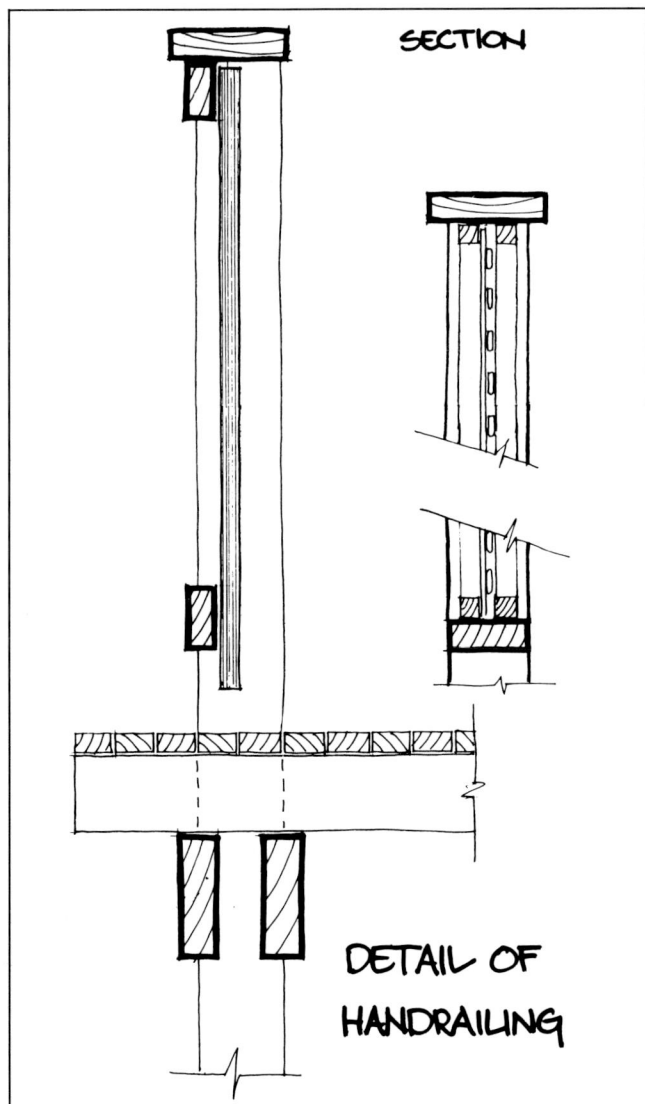

SECTION

DETAIL OF HANDRAILING

STRINGER 200×50

TREADS 300×50
HARDWOOD

BALUSTRADE

RAIL 100×38

BALUSTER 38×3?

STEEL BOLTS IN
COUNTERBORED HOLES

GROUND LINE

SPECIAL 'TEE' BRACKET
GALVANISED

HOLDING DOWN BOLTS

TREADS 300×50

MASS CONCRETE
FOOTING

STRINGER 300×50
HOUSED FOR TREADS

10MM DIA BOLT.

STAIR DETAILS

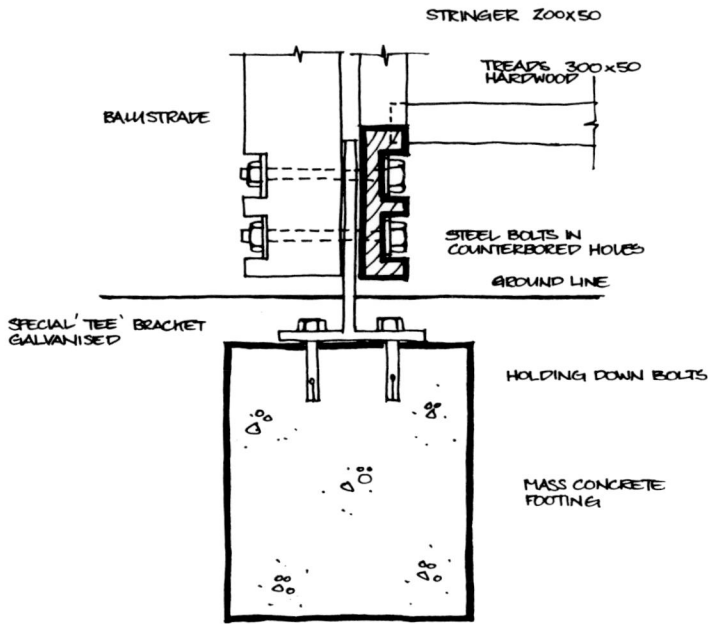

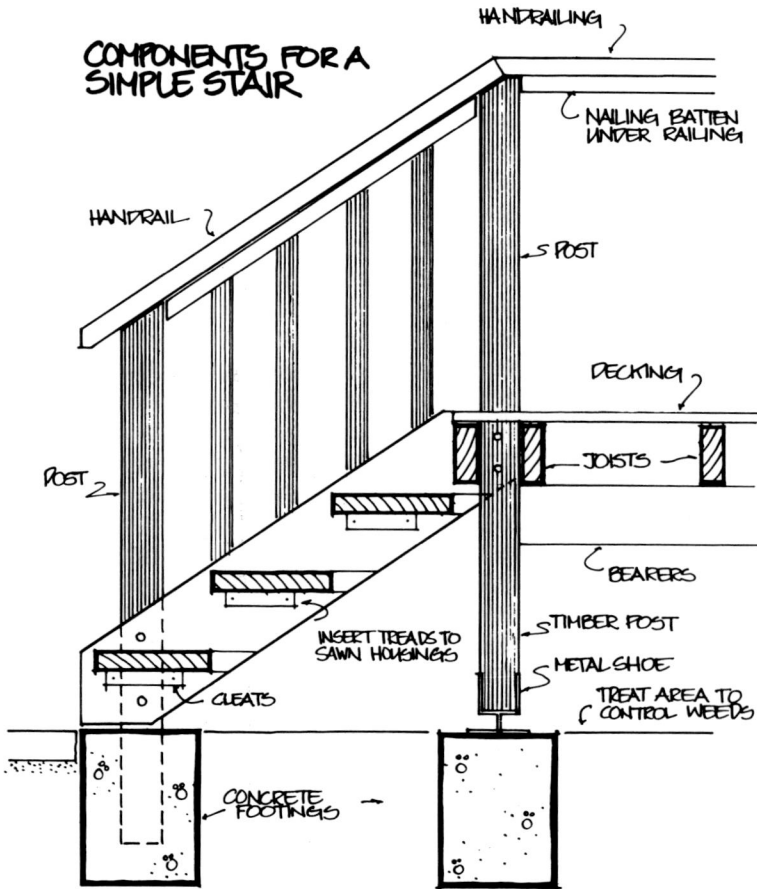

**FOOTING DETAIL FOR
STAIR STRINGER**

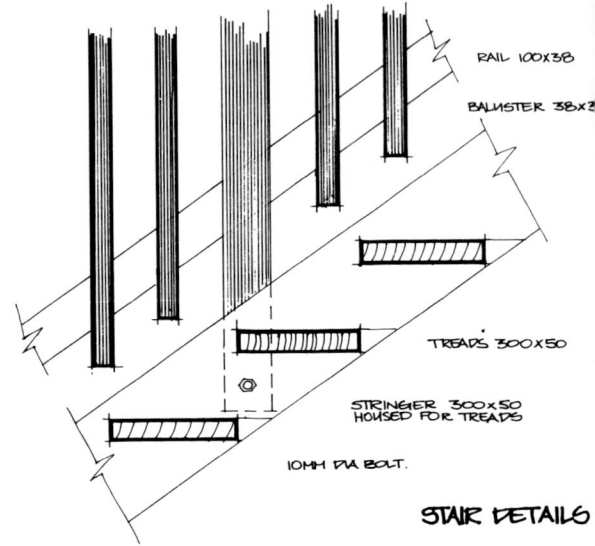

COMPONENTS FOR A SIMPLE STAIR

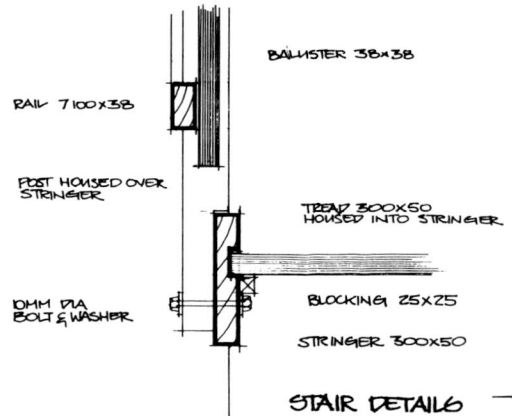

HANDRAILING

NAILING BATTEN
UNDER RAILING

BALUSTER 38×38

RAIL 7 100×38

HANDRAIL

POST HOUSED OVER
STRINGER

POST

TREAD 300×50
HOUSED INTO STRINGER

DECKING

10MM DIA
BOLT & WASHER

BLOCKING 25×25

JOISTS

STRINGER 300×50

POST

BEARERS

STAIR DETAILS

INSERT TREADS TO
SAWN HOUSINGS

TIMBER POST

METAL SHOE

TREAT AREA TO
CONTROL WEEDS

CLEATS

A simple timber stair should have treads about
300mm deep with a rise of around 175mm.
Hardwood is best for the stair treads to resist
wear, while selected timbers can be used for
other parts. Stairs require careful setting out
but the work looks more difficult than it really
is and most amateur carpenters will be able to
achieve good results using these simple
details as a guide.

CONCRETE
FOOTINGS

Railings and Stairs

On any deck that is more than 300 mm above ground, a balustrade or handrail is essential. A common method is to fix the handrail to the tops of the posts which are left longer for this purpose. These posts can be extended to support a pergola over the top of the deck as well.

Most decks need a stair to provide access to the rest of the garden. This can be simply constructed by extending the bearers at an appropriate angle to the ground to finish on a concrete pad. The treads of the stair will be housed into the sloping bearer (called a stringer).

In most timber decks, there is no need to enclose the riser (see diagrams). The handrail details of the deck will be repeated in the stairs, but take care not to leave gaps big enough for a child to fall through.

Attractive balustrades may be made with trellis or lattice work as infills. But again, remember to child-proof your work by ensuring no gap is wider than 100 mm. Another infill design includes vertical 38 mm × 38 mm dressed timber battens, nailed to the top and bottom rails.

DOUBLE JOISTS SUPPORT DECKING AT HANDRAIL POSTS

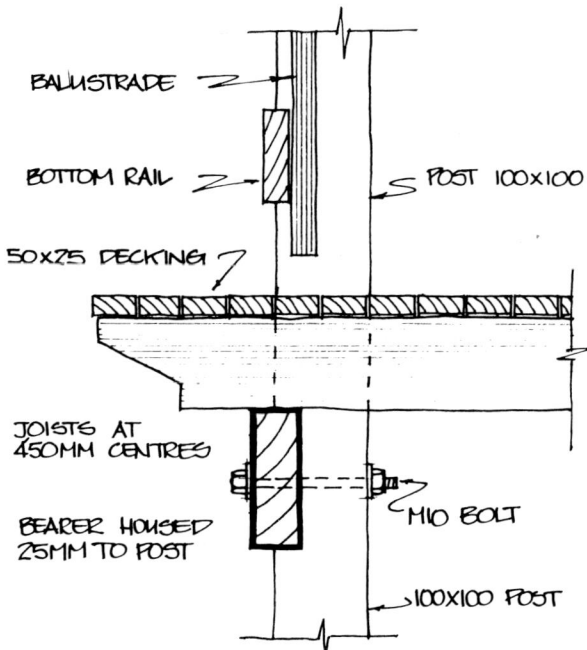

BALUSTRADE

BOTTOM RAIL

POST 100x100

50x25 DECKING

JOISTS AT 450MM CENTRES

M10 BOLT

BEARER HOUSED 25MM TO POST

100x100 POST

DETAIL OF HANDRAIL POST CONNECTION AT DECK LEVEL

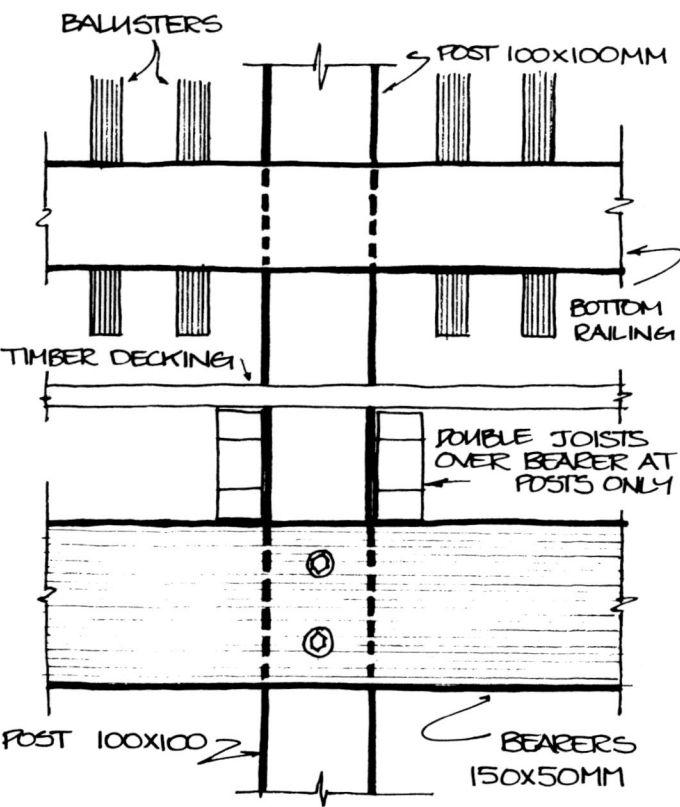

BALUSTERS

POST 100x100MM

BOTTOM RAILING

TIMBER DECKING

DOUBLE JOISTS OVER BEARER AT POSTS ONLY

POST 100X100

BEARERS 150x50MM

SIMPLE STAIR DESIGN

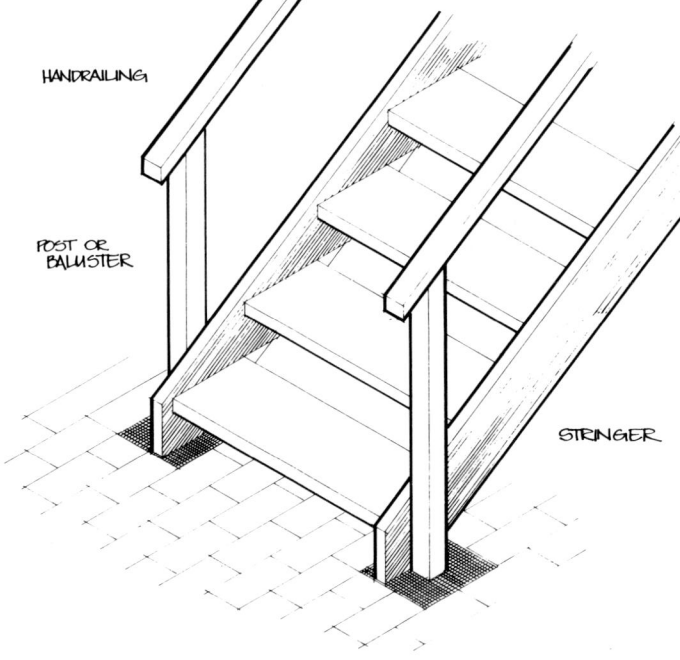

HANDRAILING

POST OR BALUSTER

STRINGER

7

FINISHING TOUCHES

Outdoor Furniture • Plants • Lighting • The Insect Problem

A constant emphasis throughout this book is on designing a barbecue area as a total concept. A barbecue should not stand alone. It should be a part of the whole garden area, complementing and enhancing the total overall landscaping of your outdoor living area.

In the earlier chapters, we have described the design and construction of the main components of a successful barbecue and outdoor living area. However, there are a number of extra items that contribute to the easy functioning and attractive look of any outdoor space. Here are the additional image-makers that will help put the finishing touches to your new barbecue area.

Outdoor Furniture

Most of the meals you prepare on your barbecue will be served outdoors. It's therefore essential that you have sturdy, weather-proof, easily maintained furniture so that everyone can relax and enjoy all the benefits of eating informally.

Australia has a climate that is extremely harsh to timber used outdoors plus a whole range of insects that do enormous damage to unprotected timbers. Western red cedar is well suited to outdoor use as it has a high resistance to rotting and insect attack although it is sometimes a little fragile and does not take hard wear well.

Australian hardwoods are also suited to outdoor furniture, especially if built-in. Recently developed timber treatments, using copper-based solutions, are also effective. However, if you are using timber treated with copper or arsenic, don't use the offcuts as barbecue firewood!

Built-in seating and tables make an attractive alternative to portable furniture if you have the skill to design and make your own. We have provided some designs for inspiration and all the basic rules still apply. Care in selecting materials and building all parts is essential because of the rain and shine element.

If you buy outdoor settings, remember that lighter grades of timber not only wear out faster but often have inferior types of fixings. Don't expect a light timber table with fragile staples to hold together in the great outdoors. The rule is to buy the best quality you can afford and take care of it with regular maintenance.

It's always a good idea to give timbered, outdoor furniture — whether you built it or bought it — a coat of stain at the end of summer to protect it from decay and insect attack through the winter.

Plants

A cool, green atmosphere to offset your tile and paving areas is a must. Planters, especially when combined with seating are an effective way of dressing up the barbecue area with greenery and will greatly add to your enjoyment of your new outdoor space.

Vegetation near your barbecue should be carefully chosen as the cooking area can get very hot. Don't make the mistake of planting quick-growing plants close to the heat. A plant that grows vigorously will need to be frequently pruned to avoid being damaged by heat or fire.

Even a recently completed barbecue area that still has not recovered from the builder's invasion can be transformed by a couple of large-sized pot plants. For that special occasion, you may even want to hire a couple of beautiful plants to dress up the area.

Some form of shading, whether provided by greenery or artificially constructed is always essential for those with delicate skins. If your budget does not extend to a pergola at this stage, a large umbrella can be an inexpensive shade-maker. A brightly-coloured Cinzano umbrella or the now popular, larger European market type umbrellas can make your barbecue area the most attractive spot in the garden.

Lighting

Adequate lighting to allow the cook to see properly is an essential but often overlooked item. Poor lighting may well be half the reason for the burnt offerings of many barbecue cooks.

Lighting the barbecue may be as simple as a single spotlight directed towards the barbecue plate. In any case, the design should be related to the style of your home; a colonial style home will probably need different lights from a super modern hi-tech designed house.

Often the easiest place to mount lights will be on the fascia board under the gutter of the house. Access to the house lighting circuits is simple and mounting the fitting just a matter of drilling a few holes and tightening a couple of screws. Make sure that the beam of light will not be blocked out by growing plants.

Don't forget the outdoor eating area. Adequate lighting helps the preparation and serving of perfectly cooked food and hungry customers like to see what they are eating.

When choosing external light fittings, make sure they are well designed and capable of working in all weathers. Take a special note of the design of the sealing systems and the ease of changing the light globes. If the globes are a special design, it's a good idea to have a couple of spares on hand. A blown globe at a crucial time could ruin a special function.

Your choice of external lighting may be limited by the amount of work the electrician will have to do to

get the wiring to the site. Where the electrician's cost is prohibitive, portable lights may be the best solution.

Regrettably, bright lights attract insects like flies and mosquitoes. One trick is to locate the light source as far away as possible. Insects are attracted to the brightest light. If the lighting is coming from spots and floodlights mounted on the fascias of the house roof, the moths and mosquitoes may not be quite so attracted to your plates, the cooking area and your bare legs. The cook's lighting should not be located near face level for the same reasons.

The Insect Problem

Some people use insecticides and electrical insect killers to solve the insect problem; others feel that these devices attract more insects than they kill and tend to eliminate the natural predators of the really nasty ones. Should you choose a 'zapper' of some kind, it is better to mount it some distance from the table so that the sound of insects being electrocuted doesn't spoil your appetite and the insects are attracted away from the food. You may find that insect repellent coils, candles or wands are the best solution.

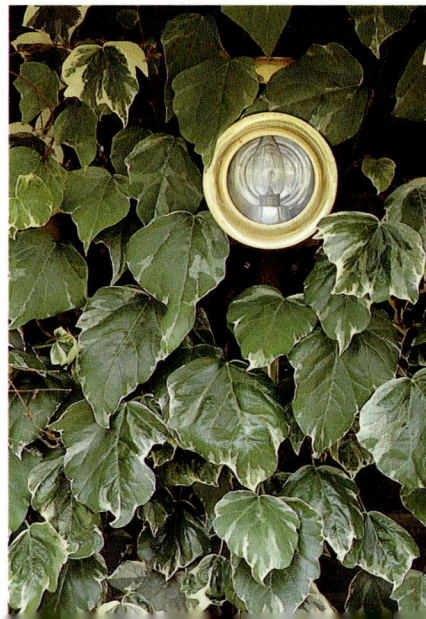

Concealed lighting can be most effective.

8
BARBECUE LIFESTYLE

Cooking Tips • Meats • Fish and Seafood • Kebabs and Shashliks • Vegetables and Salads • Spit Roasting • Safety Tips • The One Match Fire

No barbecue can create culinary miracles. Tough meat will simply become tough barbecued meat when cooked. Good quality ingredients are essential for good results in all cooking.

Preparation is the key to successful barbecue entertaining. Having all the necessary ingredients at hand, the tools close by and all the marinades, kebabs, meat, sausages and other food ready before the fire is lit will make entertaining that much easier. Also, if you haven't built in a preparation bench, set up a working table close to the fire to save yourself many trips back to the kitchen.

Make sure too that there's enough fuel. It's both frustrating and dangerous for the cook to leave the barbecue unattended at any time, especially if children are about. If you're using gas from gas cylinders, make sure that the cylinder is full or that a spare one is at hand.

If you are using wood, charcoal or barbecue beads, the fire should be glowing embers or coals with as little flame as possible. You will not want to attend to a dying fire during cooking, nor is it desirable to try and control a fire that is flaring up excessively.

Cooking Tips

Contrary to the opinion of many barbecue cooks, most people do not like to eat charcoal. If you have spent good money buying quality meat, it's a shame to destroy prime cuts by over-cooking. While there are some who like their steak to have the texture of shoe leather, you will get more compliments if your meal is cooked to perfection.

☐ To prevent food sticking, make sure you oil the grill or hotplate, also add a little oil to anything cooked in foil. Food should be wrapped on the shiny side of foil, allowing the dull side to face the heat.

☐ Remember, cooking is more efficiently and evenly done over a bed of glowing coals or embers, not a roaring fire.

☐ Flare-ups can be prevented by trimming the excess fat from the meat, then turning it when necessary to prevent fat or basting oil dripping onto the fire.

☐ The secret of good barbecue cooking is to retain the meat's natural juices. Therefore, always use tongs to turn the meat, never pierce it with a fork.

☐ Remember too, the success of your barbecue depends on the quality of the food you choose to cook. Never buy inferior cuts or use leftovers, expecting the barbecue to work miracles.

Marinating

It's surprising how much better barbecued food can taste if it has been marinated first. Used to flavour, enrich and tenderise meat, fish and vegetables, marinades will certainly add to your guests' enjoyment of their meal and enhance your reputation as a chef.

Meat can be tenderised with an acidic marinade based on wine, citrus juices, wine vinegar or yoghurt with the addition of a little oil, herbs and spices. Meat marinated in this way can be kept in a refrigerator for up to 48 hours.

A mixture of herbs and spices in oil is an excellent marinade for meat, fish and vegetables. It adds flavour as well as protecting and enriching the food. Fish and shellfish treated with this mixture need only be marinated for an hour at room temperature or kept in a refrigerator for up to 2 hours.

A delicious paste marinade can be prepared from herbs and spices, blended with a little oil, plus garlic or lemon peel. Because a paste marinade clings to the food as it cooks, it keeps it wonderfully moist.

There's also the choice of a dry marinade made from any number of herbs and spices, blended with salt. The brine, formed by the salt, is absorbed by the meat and carries with it the aromatic flavour of the herbs and spices.

Butters and Sauces

Savoury sauces and butters are delicious served with sizzling steaks and vegetables. There are many easy recipes to suit your taste buds and the quantity you require. All leftover sauce can be stored in an airtight container and kept in the refrigerator for later use.

Some sauces are best served warm, so keep them in a pot, sitting on the barbecue hotplate. Just before serving, spread liberally on the food, then place the remainder in a bowl or gravy boat for everyone's use.

Savoury butters are great favourites. Made with garlic, basil, anchovies, rosemary, tarragon, horseradish and many other ingredients, savoury butter adds a special piquancy to meat and other barbecued food. Many herb butters are also sensational served with fresh, crusty French bread.

Savoury butters are great favourites.

Sausages are always popular. Why not stuff them or make some sausage kebabs?

Meats

Almost any good quality meat will taste better if cooked on the barbecue. Steaks, chops, cutlets and even a whole fillet of pork can be cooked successfully, provided the fire is kept under control and the meat properly prepared.

It is a myth that barbecue cooking can transform cheap, tough meat into a gourmet's delight. You should always try to get first grade meat for your barbecue.

All cuts of meat are prepared basically the same way. Excessive fat should be cut off; it's the fat which causes flare-ups and an uncontrollable fire. If your guests prefer some of the fat left on, it should be cut at about 50 mm intervals otherwise it will curl and distort the shape of the meat while it cooks.

Salt as a seasoning should be avoided as it brings the juices of the meat to the surface, leaving it tough and dry. Keeping the juices in the meat also means not scoring the meat with a knife. Some cooks do this in the mistaken belief that they are letting the heat penetrate inside the meat. Also, turn the meat on the barbecue with tongs or a flat spatula, never a fork.

All meat should be taken from the refrigerator long enough prior to cooking so that it is completely thawed out and has reached room temperature. You can't expect good results from trying to cook half-frozen meat on the barbecue.

Remember barbecue cooking is fast cooking. A moment's inattention, especially if the guests have ordered rare steaks, will give a disappointing result.

Meat should be left alone on the barbecue, only moving it if the fire becomes too hot in that place, until the fat and juices begin to show on the upper surface. The meat should be immediately turned, and the same effect of the juices appearing on the first cooked side, now on top, will signal that the meat is ready to serve. Even a short delay will cause a noticeable difference to the meat.

For well-done meat, use the trick of cooking different thicknesses according to style or taste, or alter the cooking times to suit. It is impossible to reliably cook thinner steaks for a rare finish.

Chops and cutlets, especially the small lamb cutlets, are terrific cooked on the barbecue and the method outlined is still useful. Always over-estimate the amount when you cook cutlets; they're so popular it's amazing how many hungry people can eat. After all, a request for a second helping is the best compliment for the cook.

A Special Word about Steak

A well-cooked steak will have a well-done exterior no matter what the inside has been ordered to be. Mod-erately slow cooking, rather than cooking over a vigorous fire, will cook the whole of the meat. Steak that is almost burnt on the outside but raw on the inside will please few people.

For cooking steaks, the barbecue plate should be quite hot. Steaks need to be turned only once during cooking. Too frequent handling will disperse the juices and dry out the meat. Steak should be placed on the hotter part of the grill and seared. The lower surface will be sealed to lock in the juices.

There is no such cut as barbecue steak. The best barbecue steaks are T-bone, sirloin, rump, fillet and ribs eye.

Ask the butcher to slice some steaks thin, some medium and some quite thick. In that way, the outside of the meat will be perfect, the middle as ordered and they will all arrive at the table at the same time. Keeping track of each order then remains easy: the thickest will be rare, and the others medium or well-done.

Cut steak to different thicknesses to serve well done, medium or rare as requested.

Carpetbag Steak

Mince and the Hamburger

As with steaks, the best burgers are made with prime quality meat. Ask your butcher for a top cut of beef and have him mince it for you.

When preparing the mince, mix in egg and breadcrumbs for binding and add your favourite seasonings. Burgers can be prepared well in advance, placed between waxed paper and stored in the refrigerator.

Ideally, all burgers should be roughly the same size so they will cook evenly and everyone can be served at the same time.

Burgers should be cooked over a medium to hot fire for 10–15 minutes and turned half way through. Just before the end of cooking, is the best time to brush on your favourite sauce. Then, serve with rolls or buns.

Most people like hamburgers served with lots of tossed salads, mustard, pickles, mayonnaise or special hamburger sauce.

Pork

Most boned cuts of pork are ideal for barbecuing. Fillets are not cheap but there's no waste and the meat is tender. Butterfly or medallion steaks barbecue well but make sure they are not overdone.

Pork is ideal for kebabs, especially if marinated in a sweet or spicy marinade first. A boned leg of pork is a good choice for kebabs; this gives you lots of meat with very little wastage. Cooking time is about 15–20 minutes.

Ham steaks are also welcome on the barbecue, served with their ever popular partner — pineapple. Ham steaks cook in minutes as they only need to be heated through.

Any pork dish goes well with fruit flavours. Try an apricot base for a marinade or plum and chilli, red currant, a plain plum sauce or sweet and sour sauce.

Pineapple Porkers — pork chops barbecued and served with colourful pineapple and capsicum rings.

Lamb

Don't stop at the traditional lamb chop; many cuts of lamb are delicious cooked on the barbecue. A boned leg, a rack, loin or spare ribs can make a spectacular impression when barbecued.

Lamb is also delicious cubed and served as kebabs or satays. Mint, rosemary and garlic are the most popular seasonings with lamb, either as herbs or mixed in sauces. Other mouth-watering alternatives include apricot baste, green butter and oriental sauce.

The loin and rack of lamb handle easily when placed in a greased wire basket, then cooked over a good, hot fire. Always remove the skin and excess fat first.

A leg of lamb may be boned and laid out flat, wrapped in foil or stuffed and roasted on the spit.

Sausages

Most barbecue cooks have had the experience of exploding sausages. Here are a few tips to help avoid this problem. All sausages destined for the barbecue should be taken from the refrigerator at least three hours before cooking, to thaw out and reach room temperature.

Some barbecue chefs prefer to parboil sausages before cooking. This starts the cooking process and helps avoid under-cooked centres. To parboil, place the sausages in a pan of water and slowly bring them to a simmer. Now remove pan from the heat, cover and let stand for 15–20 minutes. Drain and your sausages are ready for the barbecue.

Arguments rage over whether to prick or not to prick sausages. Purists argue that by pricking holes in the skin of sausages, precious juices are lost and they become dry. Others are unable to handle un-pricked sausages without trouble.

The secret of success in cooking sausages on the barbecue is to cook them slowly over a well-controlled fire. Initially, slow cooking over the plate, followed by browning over the grill will give the best results.

Dampening a charcoal fire with an occasional spray of water from a squeeze bottle or trigger action container will help control flare-ups. Apart from beef and pork sausages in thick and thin styles, many continental sausages are perfect for barbecuing.

Try thin slices of 10 cm diameter Hungarian salami as an appetiser. This is always well received by children. Frankfurts too can be sliced in half and cooked on the barbecue.

Shish Kebabs

Spice that Sausage Sizzle with Variety

Many savoury sausages are now available and are excellent for barbecuing. Try some of the following:

Beef and Blackbean	Kransky
Beef and Tomato	Lamb Provencale
Blackbean and Honey	London Pork
Boerwors	Merguez
Bratwurst	Mexicana
Chicken	Minted Lamb
Chicken and Tarragon	Peppercorn
Chorizo	Pork and Veal
Colonial Beef	Satay Beef
Curry	Tomato and Onion
Garlic	Turkey and Chives
Italian Spicy	Turkey and Sage
Italian Tomato	

A barbecue is a great way to whip up a wonderful children's party.

Tarragon Tempters — juicy drumsticks served with savoury tarragon flavoured butter.

Poultry

Chicken, duck and even turkey can be cooked on the barbecue, either whole, in portions or roasted on a spit (see Spit Roasting). Some larger birds may be partially cooked in the microwave oven, remembering that microwaves cook from the inside first, in contrast to the barbecue. This partial cooking will dramatically reduce the time on the barbecue.

The best way to cook a whole bird is to break the backbone so that it will lie flat on the barbecue. Cook over a slow heat on the plate or grill. Chicken is particularly well suited to marinades and basting. Care should be taken during cooking to baste the bird regularly to prevent the meat drying out.

Wire baskets which hold a whole bird are useful when cooking poultry. The baskets are available in a variety of sizes, right up to ones capable of holding a large, whole fish. Turning a larger bird and working with it on the barbecue is made much easier and the purchase of a couple of wire barbecue baskets is recommended.

Fillets from the breast or thigh of the bird can be cut into bite-sized chunks and served as kebabs or satays. Always remove the skin first, cook over a medium fire and baste just before serving. Cooking time is about 15 minutes.

Fish is easily barbecued in special wire baskets.

Fish and Seafood

Many barbecue cooks are afraid to try and cook seafood on the barbecue. As with other foods, the key to success is careful control of the fire to ensure the fish is not over-cooked.

Whole fish, fillets, steaks or cutlets, crayfish, lobsters, yabbies, bugs and even crabs as well as prawns and shellfish taste great when barbecued. Some seafoods are at their best when grilled over an open fire.

For special occasions, your local fish shop will be able to supply larger fish if you order in advance. Specially designed wire baskets make cooking any sized whole fish simple. The troublesome job of turning the fish is suddenly made easy.

The fish should be thoroughly cleaned, scaled and gutted but rinsing in water should be kept to an absolute minimum so that flavour is not lost. The skin should be lightly scored with a knife, but not too deeply. For presentation some cooks trim the fins and the tail.

The fish is then clamped in a lightly-greased wire basket ready for cooking. Fish should never be served undercooked and a meat thermometer is an excellent investment if you intend to cook a large fish on the barbecue. When the internal temperature has reached 65°C, your fish will be done to perfection.

Fish fillets are rather delicate to handle and best cooked in aluminium foil. Use only the premium grade of foil and use enough layers to make sure the fish may be safely turned. Use a little oil to coat the inside of the foil and add a knob of butter and season to taste. Place food on the shiny side so that the dull side faces the heat source.

Depending on the thickness of your fillets, they will take between 10 and 15 minutes on a slow to moderate fire. Turn the foil-wrapped fish fillets several times during the cooking. If the fire is too strong, use the dampening technique to control the intensity. Dampening will also help to add precious moisture and stop the fish drying out.

Fish cutlets and steaks can be cooked on the barbecue plate and make a great change from filleted fish. Ask your fish shop for good-sized cutlets and make sure they are reasonably thick. Fish cutlets are more easily handled on the barbecue.

Cutlets and steaks cook quite quickly, so a slow to moderate fire is suggested. Prepare the barbecue plate with a little oil or butter to stop sticking. Fish requires little preparation and any seasoning may be added on the barbecue as it is cooking. Fish is fragile and should not be handled too much. Turn the cutlets only once, otherwise you may risk disintegration.

Crabs, lobsters, crayfish, yabbies and bugs are best cooked in their shells if a good size. Cook them with the shell down for most of the time to keep the meat moist and give them a final burst on the flame to add colour and taste. A large lobster takes about 30 minutes to cook over a moderate flame. Dampening the fire with the occasional spray of water will help preserve moisture.

Prawns can be barbecued on the plate with barbecue sauce or in butter. Shelled, green prawns may be cooked on skewers but leave the tails on. Barbecued prawns can be served with a variety of dips and sauces for extra effect and are also a wonderful appetiser for a fish main course.

Kebabs and Shashliks

These two terms are interchangeable for meat, poultry, seafood and vegetables cooked on a skewer. Skewers should be at least 20 cm to 30 cm long as you need to leave 2 cm to 3 cm at the end for ease of handling. Choose flat skewers — the food stays in place when you turn them.

The chief attraction of kebab-style cooking is the variety. You may even want to prepare the ingredients and then let your guests make up their own. Longer skewers are available, up to 45 cm long which feed two or more people. Combinations of meat, vegetables, fruit and marinade can make a delicious taste sensation on a skewer. However, some foods require longer cooking than others, so your combinations should be carefully thought out. Also, if you combine small pieces of one meat with other larger pieces, it will be difficult to tell when they are ready. A mistake is often made by trying to get too much meat on one skewer. Enough room should be left between pieces of meat so that they can be cooked all round.

You don't want to turn the skewers by hand.

A variety of vegetables creates colourful kebabs.

Specialist barbecue supply shops have a variety of frames and kebab turners, both manually operated and motor driven. The motor driven style will save effort and gives an evenly cooked result. The frames are a great idea as they lift the skewers from the grill or plate and allow even cooking above the flames.

Satays

Satays can serve as useful snacks or appetisers at a barbecue. They originated in Asia and are becoming increasingly popular because of the great variety of different recipes.

Satays have two parts: a marinade and a sauce which is served with the food. While the preparation may be tedious, they are easy to cook and take only 10 minutes or so before they are ready. For a larger party where the satays are served as an entree, it may be a good idea to cook them on a separate small portable barbecue, keeping your main barbecue free for the main meal.

Satays are intended to be eaten directly from the skewer, and you will be amazed at the amount a hungry person will be able to eat. Beef, pork, lamb and chicken are all suitable for satays and the meat should be prepared at least 12 hours in advance so that the marinades can do their work.

First, thread about 6–8 small cubes of meat onto the sharp end of the skewer and leave a generous handle. The cubes should be about 15 mm in size. Allow the meat to stand in the marinade for at least 12 hours but keep refrigerated. Cooking satays is best done on a hot fire while lightly basting them with the remaining marinade or oil.

Serve the satays with a variety of sauces so that your guests may pick their favourite one. Soaking bamboo satay skewers in water for a few hours will help prevent them being burnt on the fire. A properly set fire will have little flame, so burning the wooden skewers should not be a problem.

Vegetables and Salads

A successful barbecue doesn't end with sizzling steaks and sausages, tasty chicken or fish. Combining these delicious meats with just the right vegetables and salads makes all the difference.

Most vegetables can be cooked on the barbecue wrapped in foil, although make sure you use a heavy-duty foil or double wrapping to prevent tearing.

Vegetables are also attractive sliced into bite-sized pieces and served as kebabs with creamy bechamel, hollandaise or bearnaise sauces.

Who can resist potatoes baked in foil and topped with generous lashings of sour cream and chives? Other barbecue favourites are corn on the cob, stuffed tomatoes and barbecued mushrooms.

Tossed green salad, coleslaw, potato salad, an eye-catching Greek salad with black olives and feta cheese, a Waldorf or tabouli salad — they're all just perfect for barbecues.

Spit Roasting

Specialist barbecue suppliers sell attachments that convert almost any home barbecue into a spit roast or rotisserie. These accessories are quite cheap, but if you plan to use them frequently you should consider buying a robust machine. There are battery and electrically operated models as well as manual ones. Spit roasting accessories can also be hired for special occasions.

Even a small chicken will take around one hour to cook on a rotisserie and the cost of an electric model will be quickly appreciated by the cook. The key to successful spit roasting is the slow and constant rotation of the food so that the juices are constantly basting the food as it cooks. Hand rotation rapidly becomes tiresome and tedious.

A well-designed rotisserie will have substantial prongs to hold the food and ensure that it continues to rotate throughout the cooking. Some of the better models will also supply a set of clamps which help to secure the food.

Spit Roasting Poultry

To prepare a bird for spit roasting, the neck should be removed and the skin and cavities well rinsed and patted dry. The bird should be secured with string or wire so that the wings are flattened against the breast and will not shake free as the spit rotates.

Also tie the drumsticks and the tail together, grease the shaft of the spit, slide one set of prongs onto the shaft, facing away from the handle and push the point of the shaft through the neck. The two sets of prongs will oppose each other and the shaft should be preferably a square section.

Place the bird in the middle of the shaft and test for balance. Your spit motor will be unnecessarily stressed if the balance is wrong and the extremities may become loose. Only when the balance is right is the spit ready. Lightly baste the skin with butter, oil or other recipe and the cooking is ready to begin.

As a very rough guide, allow about 45 minutes per kilogram for cooking, but the only way of making certain that the meat is done is to use a reliable meat thermometer inserted into the breast of the bird. When the temperature of the deeper meat has reached about 85°C, the bird will be properly cooked. Smaller birds can be tested simply by piercing the breast. If the juices are clear and the joints move freely, the meat will be ready.

Naturally any cooking times will be learned by experience. Each barbecue is different and the intensity of your fire, its efficiency and other factors will influence the final result. Experiment on your own family before risking a new technique on an important occasion.

A larger turkey takes as long as 5 hours to cook and you may well have to be prepared to entertain your guests for quite some time before they eat. The essential part of all spit roasting is care and attention to the fire to ensure a constant and moderate heat.

A gas barbecue gives the best results in spit roasting although wood, charcoal and other fuels may be used as long as you attend the fire constantly to keep the cooking under control. The use of reflectors and drip trays to increase the efficiency of cooking and so reduce cooking times is suggested.

Salads are ideal fare when dining out of doors.

For that special occasion with many guests, nothing can beat spit roasting.

Spit Roasting a Whole Lamb or Pig

For that really special occasion or a big party, nothing can beat spit roasting a whole animal for excitement and special effect. A whole pig or lamb weighs around 30–40 kg which is a lot of meat.

You will need to specially order the carcass from your butcher and his assistance and advice can be very useful. Very few homes will be able to store and refrigerate such an amount of meat so it's best if you pick up the carcass from the butcher on the morning of your party.

While the preparation of a whole beast for spit roasting is a considerable amount of work, the cooking is relatively straightforward. Given the weight of the carcass, the equipment needs to be substantial and can be hired from a number of companies. A gas-fired spit roast will give the best results and is the simplest to control. Depending on the equipment and the skill of the cook, a full carcass usually requires some 6–8 hours to cook.

The full carcass should be thoroughly defrosted and allowed to reach room temperature. Some experts recommend that the carcass be placed in a bath of very hot water for 30 minutes or so to pre-heat the meat, convert some of the fat and so reduce the cooking time. Allowing the meat to reach room temperature should be adequate if pre-heating is impossible.

Placing the carcass on the spit is the critical part of the whole exercise. If this is not done properly, the whole animal may fall from the spit into the fire. Use a butcher's knife to cut all the sinews that connect each rib inside the cavity. Slit the thick meat around the thigh and under the front legs to allow the heat to penetrate.

As each spit manufacturer uses a different design, read the instructions carefully and follow them to the letter. With the carcass on its back, the spit is inserted from the rear, close to the backbone and pushed through to emerge through the throat. Prongs are then inserted from both ends and the legs stretched out and secured with strong wire.

Some better spit systems have special clamps and leg supports to assist in securing the carcass. All extremities should be tied with wire. Balance is critical to avoid overstressing the motor. The unit will be running for between 6–9 hours, so care and attention to detail at this stage is critical.

Controlling the Fire

Gas is the best fuel for spit roasting and it is fast and controllable. Wood, solid barbecue fuel and charcoal may be used, but the fire will need constant attention and regular replenishment. If you can borrow or hire a gas-fired unit, this type is best for the inexperienced barbecue cook.

Fat collection trays are essential. A whole pig and, to a lesser extent, a lamb will release a substantial amount of fat which represents a potential danger if it is accidentally ignited. You will need to keep an eye on the spit and empty the collection trays regularly.

90

Meat Thermometers

Absolutely essential when spit roasting a whole pig or lamb, a meat thermometer should be inserted at an angle so it does not fall out as the meat turns.

It should be inserted in such a way that it does not touch bone, fat or the spit. A meat thermometer is the only way that you can be certain that the meat is properly cooked when you begin to carve the carcass. In the thickest part of the meat, the temperature should reach between 80°C and 85°C before serving.

Carving the Meat

Carving at the spit and serving the mouth-watering meat will be the culmination and climax of the cooking. Many cooks are afraid of this job but there is no reason to be timid. If you have taken care of your work during the preparation and cooking, you'll be confident about carving.

Don't attempt to cut individual serves from the carcass on the spit. Cut generous chunks from the spit and remove them to the table to be sliced into individual serves. Keep the spit revolving to complete the cooking of the meat inside for second helpings. Usually at this stage of the party, informality has taken over and many guests will want to help themselves.

For sheer excitement at an outdoor barbecue, there's little to beat spit roasting a whole animal and the event is well worth a try for that really special occasion.

Safety Tips

Any situation outdoors where there is heat, an open flame, fat and a collection of people is potentially dangerous. First, there is the hazard associated with lighting the fire. It can be annoying and downright frustrating when a crowd of hungry people are standing around and the charcoal is damp, the wood soggy and no amount of matches seems to get the fire going.

Herein lies the temptation to throw a bit of lighter fluid, kerosene or petrol on the fire. **Never be tempted**. Using flammable substances to light a barbecue could not only land you or your guests in hospital, it could be lethal.

The use of flammable liquids to start fires is dangerous, irresponsible and downright stupid. Any flammable or explosive substances or liquids should be well removed from the area of the barbecue. Each year, many people including adults and unfortunately children, end up in hospital with shocking burns as a result of attempting to start backyard fires or barbecues with explosive liquids.

There are also the dangers associated with occasional flare-ups from the fire itself. This often happens when fat from the meat has dropped into the flames. Because it's usually quite unexpected, a sudden surge of flame will burn the cook's hands and arms and could possibly ignite a portion of clothing, such as a plastic apron.

Therefore, always keep a squeeze bottle of water at the barbecue and at arm's length. Water can be used quickly to douse any unwanted, rising flames and also used to control the heat you need for cooking. It's also a good idea as an emergency back-up to have a bucket of water and a hose close by.

Smoke is also another hazard which can make things unpleasant for guests, especially those with an allergy, and can often lodge an irritating piece of grit in someone's eye. Also watch for sparks, particularly in dry weather and on windy days.

It's not always possible to control smoke but a well-designed barbecue, one which takes into account the prevailing winds, makes sure the offending smoke heads in the right direction, away from guests. Perhaps eye drops won't go astray in your first aid kit.

Remember, it's the cook's responsibility to see that all safety measures are in place before proceedings start and once the cooking has commenced, never leave the grill and hotplate unattended.

A moment's distraction could find children picking up tools, playing with the fire and a tragic accident occurring. Make sure there's enough space for cooking and serving the food. This is one way to minimise the likelihood of any mishap taking place.

If you are barbecuing in the bush, make sure you check with local authorities on the fire regulations. There could be a total fire ban in force. And most importantly, when the party is over, clean up all the mess and put the fire out completely.

The One Match Fire

This is a simple, tried and tested fire-starting method that in ten years of lighting barbecues has never required more than one match. No petrol, kerosene or other dangerous methods should ever be contemplated. This one match method of starting a fire is simple and safe and requires just two sheets of newspaper and a small supply of kindling.

Tear up the newspaper into strips and roll them up into a loose ball. The kindling (small strips of dry softwood) is laid around the newspaper to form a teepee or tent-like shape, taking care to leave enough room for airflow. The cook should have at hand a supply of wood about 25 mm square by about 300 mm long, ranging up to the largest pieces of some 75 mm square by 300 mm long.

Simply light one match and use it to light the paper at the base. As the fire takes hold, the smaller timber is added until the fire is well alight. Care should be taken not to choke the fire by adding too much, too quickly. Dry hardwood is the best fuel to use once the fire is alight as it gives off less smoke and burns more slowly.

9

BUYERS' GUIDE

Which Type of Barbecue? • Accessories • Utensils • Crockery • Outdoor Furniture

A visit to a barbecue speciality store is a must for barbecue builders or buyers. These shops carry a comprehensive range of portable barbecues, accessories and utensils plus the hotplates, grills and burners you need if you are building your own. Don't hesitate to ask the sales personnel for information or advice — that's what they are there for.

Which Type of Barbecue?

For those who have no intention of building a barbecue, the range of portables on the market can seem bewildering. Basically you get what you pay for. But just because you want something simple or small enough to take on picnics doesn't mean you have to take a drop in quality. Check the finish and stability of portable barbecues before you buy. How easy will it be to clean? Does it have a hotplate and grill — or just a grill?

It is important that the cooktop can cope with the style of food — and the quantity — that you will be barbecuing. If you plan to be adventurous and try a wide range of barbecue meals, invest in a barbecue with both grill and hotplate. If you want to include roasting and steaming in your repertoire, you will also need one with a lid. One very important consideration, especially if you are going to be using solid fuel, is the distance of the grill from the heat source. Remember, it is heat that does the cooking, not flames. If the food on the grill is too close to the fuel you may have problems with fatty flareups and charred meals.

Before you buy, decide on the type of fuel you would prefer and make a realistic assessment about the number of times you will want to use the barbecue each year. If you are confident that you will make good use of the barbecue, then buy one a little larger rather than smaller. It is better to have too much cooktop than not enough and have family and friends waiting constantly at your elbow to be fed. It is also fairly common for people to become more adventurous in their cooking once they have mastered their new barbecue and possibly for the unit to become the focus of summer entertaining.

Barbecues come in all shapes and sizes. Most styles are available for cooking with either solid fuels such as charcoal and barbecue beads or with gas. You should have no problems finding one that suits both your needs and your garden to perfection. It is also important that you have a place to store your barbecue when you are not using it. For long life it is important to protect your investment from the weather.

The most basic portable barbecue consists of a pressed metal fire pan with a circular grill, usually with some method of raising the grill above the fire. A good quality model of this type will have a generous coating of porcelain enamel to the fire pan and the grill will be chrome-plated. You should look for good quality plating to ensure that your new barbecue will not rust away.

The kettle-style barbecue is increasingly popular and it offers a range of accessories that would satisfy the most demanding chef. Because the kettle barbecue operates at a very high temperature, it is important that the fire pan has a high quality finish. An inferior pan may become distorted and eventually rust through.

Next best thing to a built-in barbecue, is a portable barbecue with a food preparation area. Many of the models now available have an attractive redwood trolley that provides the cook with a little workspace. You need to look for quality here too. The difference between first-rate and second is in the thickness of the timber, the type of fixings and the overall finish.

Accessories

Barbecues come with a wide range of optional extras these days, from rotisseries to kebab turners, heat reflectors, wok holders and covers. The first accessory that a sales assistant will try to sell you is likely to be a rotisserie. There are manual models that cost practically nothing, ranging to sophisticated models with variable speed, electric drives.

If you intend to use this feature, you should invest in a motor-driven model. Turning the spit manually for an hour or so to cook even the smallest bird will quickly become an unwanted chore. Remember, a reasonably sized joint of meat, enough to feed six people or so, will weigh at least a couple of kilograms. If the meat is out of balance on the spit, very considerable strain will be placed on the motor of the unit.

Battery operated models are suitable for light to medium use. Make sure to have a spare set of batteries on hand as they always seem to run out at the critical time. If you can afford it, the electrically operated units are the best investment and the extra money is well spent. Again, when shopping around, look for the sturdiest models.

Utensils

While it is probably fair to say that poor workers often blame their tools, it is equally true that poor quality tools make good results difficult to obtain.

Your barbecue tools should be sturdily made for they will be used often and provided you are careful, should last a lifetime. They should be long enough to use safely on a hot cooking surface and have well-insulated handles. Remember plastic handles will melt if accidentally left on the grill or hotplate.

As a general rule, you may expect to pay twice as much for a good tool as a cheap one but oftentimes the expensive item will last longer and perform better. Take particular note of the quality of any welds that occur on a tool; it's in this area that most cheap tools fail.

Barbecue utensils need long handles.

Basic Essentials

APRON A sturdy neck-to-knee apron is best to guard against splatters.

BASTING BRUSH Small 5 cm wide paint brushes are handy for dabbing bastes and sauces over food.

BRUSH A stiff metal brush is essential for scrubbing the grill.

FIRE BLANKET or FIRE EXTINGUISHER If an emergeny arises, either of these items will quickly extinguish flames.

FOIL Use heavy-duty aluminium foil for wrapping potatoes, vegetables or bread.

FORK A long-handled fork is a useful aid when turning food or pricking sausages.

KNIVES Efficient meat trimming requires a butcher's knife, while fruit and vegetable peeling is eased by a parer knife.

MEAT THERMOMETER This long probe with a temperature dial at one end is inserted into the meat or poultry before cooking. As the meat cooks, the dial indicates the internal temperature. A meat thermometer ensures there are no nasty surprises when serving — especially if you have a large piece of meat on the spit. Often it is easy to assume the meat is ready because of the external appearance, when in fact it is still uncooked inside. The thermometer eliminates error. Meat thermometers are particularly useful for spit roasting and when using a covered barbecue. They are absolutely essential if you are barbecuing a whole pig or lamb.

OVEN MITT OR CLOTH Avoid burnt fingers with a comfortable pair of oven mitts or a thick oven cloth.

SKEWERS Flat, metal skewers with an insulated handle are best. If using bamboo satay sticks, soak them in water beforehand. The hairpin style skewer is also useful and you can make it yourself. Bend stainless steel wire into hairpin-shaped skewers about 20 cm long and 1.5 cm wide. (You can make them while you watch television.)

SPATULA Foods that tend to fall apart such as hamburgers and fish need to be turned with a long-handled spatula.

TONGS Turning food and adjusting hot coals is easy with a pair of long-handled tongs.

WIRE BASKETS Keep large pieces of food like whole fish in barbecue wire baskets. You can buy these or make them yourself using galvanised steel mesh.

Crockery

In many ways it's an insult to the cook to have a well-cooked meal presented to guests on soggy paper plates. Half the enjoyment of eating a delicious barbecued steak is to have it served in style on attractive crockery.

There are so many casual, inexpensive dinner settings on the market these days that the use of paper plates in one's own garden seems inexcusable. Naturally, there's the hassle of the washing up afterwards, but the success of the whole function could rest on your presentation of the food. Why let yourself or your cook down after all the hard work you have done to satisfy the hungry hordes?

Outdoor Furniture

The outdoor furniture you choose will be largely determined by the style and architecture of your home and landscaping. There are so many manufacturers supplying outdoor furniture and accessories that the biggest problem you will experience is making a choice.

The traditional timber furniture, usually made from either western red cedar or treated pine is still the most popular but some modern materials have been introduced recently. UPVC (a specially developed UV resistant plastic) pipe framed furniture is both modern and practical and a number of manufacturers have impressive ranges to suit most tastes.

Another traditional material that has made a comeback in recent times is cast iron and the reproductions in cast aluminium. Corrosion resistance and light weight are the main attractions of cast aluminium although some purists feel that the weight of cast iron is part of that material's charm.

A recent innovation has been the introduction of engineering plastics into the outdoor furniture market. Manufacturers have overcome the problems of sunlight damage and there is an exciting range available. Plastic furniture is light and easy to handle. Many local manufacturers now compete with importers for a piece of this market with new ranges becoming available almost monthly.

A visit to the local building information centre or nearest department store or barbecue specialty shop will reveal an astonishing range of garden furniture to suit all tastes and every budget.

WHERE TO GO FOR ADVICE AND INFORMATION

New South Wales

Building Information Centre
525 Elizabeth Street
South Sydney 2012

Royal Australian Institute of Architects
55 Mountain Street
Ultimo 2007

Standards Association of Australia
Standards House
80–86 Arthur Street
North Sydney 2060

Timber Development Association
525 Elizabeth Street
South Sydney 2012

Victoria

Building Development and Display
332 Albert Street East
Melbourne 3000

Royal Australian Institute of Architects
30 Howe Crescent
South Melbourne 3205

Standards Association of Australia
Clunies Ross House
191 Royal Parade
Parkville 3052

Timber Merchants' Association
184 Whitehorse Road
Blackburn 3130

Queensland

Royal Australian Institute of Architects
Cnr Mary and Albert Streets
Brisbane 4000

Standards Association of Australia
447 Upper Edward Street
Brisbane 4000

Timber Advisory Bureau
5 Dunlop Street
Newstead 4006

South Australia

Building and Home Improvements Centre
113 Anzac Highway
Ashford 5035

Royal Australian Institute of Architects
GPO Box 2438
Adelaide 5001

Standards Association of Australia
11 Bagot Street
North Adelaide 5006

Timber Development Association
of South Australia
113 Anzac Highway
Ashford 5035

Western Australia

MBA Building Information Centre
161 Havelock Street
West Perth 6005

Royal Australian Institute of Architects
PO Box 191
West Perth 6005

Forest Products Association
of Western Australia
103 Collins Street
West Perth 6005

Standards Association of Australia
11–13 Lucknow Place
West Perth 6005

Tasmania

Royal Australian Institute of Architects
GPO Box 1139L
Manuka 2603

Standards Association of Australia
97 Murray Street
Hobart 7006

Tasmania Timber Promotion Board
68 York Street
Launceston 7250

Northern Territory

Royal Australian Institute of Architects
PO Box 1017
Darwin 5794

Standards Association of Australia
c/- Master Builders Association
191 Stuart Highway
Darwin 5790

Australian Capital Territory

Royal Australian Institute of Architects
PO Box 99
Manuka 2603

GLOSSARY OF BUILDING TERMS

AGGREGATE Crushed stone screened to a size suitable for concrete. The maximum size of stone, suitable for general use is 20 mm. Aggregate is obtained from better builders' suppliers. It is sometimes called bluemetal or simply metal. Bluemetal is crushed basalt rock.

ARCHBAR A steel structural beam designed to support brickwork over an opening such as a window or doorway. A common size of archbar is 75 mm × 10 mm flat bar. Archbars are commonly galvanised and are supplied in various lengths, according to customers' orders.

BALUSTER These are the individual members of a balustrade.

BALUSTRADE A railing usually around a deck or balcony. They may be constructed from timber or metal, but for the average handyman, timber is worked more easily.

BATTENS In general terms, a piece of timber, having a small dimension and no major structural purpose. Battens are used to support roof tiles and for shading on pergolas and are usually about 50 mm x 25 mm.

BEARER In floor construction, the structural member that carries the load of the joists to the structure. Bearers are usually fixed to posts, stumps or to brick piers.

BLEEDING A feature of unseasoned or green timber that will cause staining of concrete until all the sap leaches out of the new timber. More an unsightly nuisance than a serious defect, it may be minimised by buying well-seasoned timber or secondhand demolition timber.

BLINDING A levelling bed of sand or other fine grain material usually laid under paving.

BOLSTER A short-handled hammer with a heavy square head used for cutting bricks etc. Sometimes also called a lump hammer or a lumpy.

BRICK SETT A heavy, broad-bladed, cold chisel used by masons and bricklayers for brick cutting.

BUILDERS' SAND A clean, beach sand that has been washed to remove all salt and organic matter. Available from all builders' suppliers, either in bulk or bagged.

BUSH SAND An alternative to builders' sand but only suitable for mortars and not for rendering work. Bush sand gives a fatty mortar that is easy to work.

CEMENT Portland cement is used as a basis for all bricklaying mortars and in concrete. One bag weighs 50 kg which means there are 20 bags to 1 tonne. Cement should be purchased in sufficient quantities for the job at hand; it tends to harden with atmospheric moisture.

CEMENT MORTAR A general purpose mortar for all bricklaying. It comprises one part Portland cement and three parts sand. This mortar gives a grey to off white joint.

CHAIR A device designed to support the reinforcing steel and maintain it in the desired position during the pouring of concrete. Chairs are used in conjunction with pressed metal pans to prevent the plastic underlay from being damaged.

COMPO MORTAR A bricklaying mortar which comprises one part hydrated lime, one part Portland cement and six parts sand. A weaker mortar than 3:1 cement mortar but more economical and simpler to work.

CONCRETE A mixture of cement, sand and aggregate mixed in varying proportions, according to the strength and purpose required. For footing and sundry paving work, a mix of four parts aggregate, two parts sand and one part Portland cement is satisfactory.

CORED BRICKS An alternative to a frog in a brick. Usually found in extruded bricks, the cored out holes reduce the weight of the brick and perform the same function as the frog.

COUNTERBORING A method of carpentry which conceals the heads of bolts and nuts in posts and other timber construction. The centre of the intended bolts is marked and the first drilling is made with a boring bit, of sufficient diameter to clear the washer and deep enough to hide the bolt head. The clearance hole for the bolt shank is then drilled through. To drill the counterbore where the drill has emerged, a piece of scrap timber may be clamped over the hole and the centre carefully marked. When drilling such a hole, constantly check the depth to ensure that the counterbore is correct.

COURSE In all masonry work, a course is a rise in level of one unit.

DECKING The actual walking surface of a timber deck. Usually made from dressed and pretreated hardwoods like tallowwood.

FOOTING The lowest part of a building or structure that rests on the ground. Usually constructed from reinforced concrete or, in small structures, of bricks.

FORMWORK A timber frame made out of stakes driven into the ground and perimeter boards that sets out the level of the top edge of a concrete slab. Formwork should be well constructed so that it will not move during the concrete pour, but should also be designed that it may be easily removed when the concrete has set.

FOUNDATION The ground on which the footing is built. Sometimes used interchangeably but incorrectly with footings. This term relates to the ground not the building.

FROG A recess pressed into a clay brick before firing. The frog assists the levelling of the brick courses and strengthens the wall. (See also Cored Bricks.)

GALVANISING An electrolytic process that coats raw steel with a protective layer of zinc. This process is recommended for all exposed external steel work. Hot dip galvanising is the most effective process and users should specify this type of service.

HEADER BRICK
HEADER COURSE A header is a brick that is laid across the line of the wall of brickwork. A brick on edge header is a header course laid with bricks on edge rather than on the flat and is a useful finish to a wall.

HOUSED JOINTS A housing joint is made when a piece of timber is cut out to accept another member or when a plate is rebated to accept a stud as shown (see diagram).

JOIST A horizontal structural member, supporting a floor, deck or ceiling. The joists are fixed to plates in the case of ceiling joists or to bearers in the case of a floor. In pergolas, sometimes the terms joists and rafters are used interchangeably.

KNEE BRACE A short, diagonal bracing member usually applied to the bracing of a pergola to post connection. Bracing is essential to stop a pergola swaying.

LARRY A long-handled hoe with a hole in the blade. It is most useful for mixing mortar in a wheelbarrow.

LIME Hydrated builders' lime is supplied in 50 kg bags and is a useful additive for all bricklaying mortars as an aid to workability. (See also Compo Mortar.)

LEVEL Quite obviously, this means perfectly horizontal. If in doubt, check your construction several times with a long, spirit level, reversing the level to eliminate errors.

MERCHANTABLE GRADE TIMBER A lower quality grade of timber that will suffice for all but the most critical applications. This grade should be specified for most jobs.

PATH MESH A grade of pre-fabricated steel reinforcement designed for light paving work.

PERGOLA An unroofed or partially-roofed frame, designed to provide a base for climbing plants and to give quasi-shelter. Pergolas provide a transition space from the inside of the house to the outside and are an effective part of an outdoor living space. Pergolas are occasionally freestanding, but are usually attached to the house.

PERPENDS The vertical joints between bricks.

PLATE When applied to timber framing, this term means a horizontal load bearing member. A bottom plate may be found at the base of a stud wall and a top plate will be at the top of the wall, supporting the ceiling joists and roof structure. A wall plate is fixed to a wall to pick up the rafters of a pergola or other structure.

PLUMB A builders' term meaning perfectly vertical or perpendicular.

RAFTER An angled, supporting member, usually in a roof. In a pergola where the angle or slope of the rafters is small, the rafter may be confused with a joist.

REINFORCEMENT REBAR Steel bar stock and prefabricated mesh used to strengthen concrete work. For complicated work where suspended slabs or heavy loads will be supported, the advice of a qualified structural engineer will be necessary for the reinforcement steel design. For minor building works, such as paths and barbecue footings, the use of path mesh should suffice.

SCREED A thin layer, usually of mortar, used as a bed for tiles or a finishing application to rough concrete.

SELECT GRADE In timber, a superior grade that costs more but allows the purchaser to choose the best piece of timber for a specific purpose.

SHADECLOTHS Woven fabrics usually made from man-made fibres, either of split tape or filament. Most have especially good resistance to sunlight and UV degradation. Shadecloths are available in several grades which are specified by their light transmittance. A 70 per cent shadecloth will block out 70 per cent of the light. For pergola and shadehouse use, grades of 80 to 90 per cent are suggested.

SHOE OR CONNECTOR A prefabricated metal connection device, specially designed to fit the bases of timber posts and connect into concrete slabs and footings. Many styles are available ex-stock from hardware suppliers, and may be made to suit a particular application by a well equipped metal worker.

SOLDIER COURSE A course of bricks, often used as a finish to a wall, laid with the bricks vertical. A very elegant detail if well executed.

SPARROW PICK A technique which uses a sharp-pointed pick to roughen a concrete surface in order to obtain a better surface for other materials to adhere to.

STOP CHAMFERING The planing of the edge of a piece of timber is called chamfering. When the chamfer is stopped short of the ends or connections, it is called stop chamfering. Chamfering is desirable to reduce splitting and splintering and gives a neat, professional finish.

STRETCHER BOND Common brickwork where the vertical joints, called perpends are staggered by one half brick on each successive course.

STRINGER The angled beam that supports the treads on a stair.

STUD
STUD FRAMING The vertical timber of a wall frame. Usually the frame is from 75 mm x 50 mm or 100 mm x 50 mm timber and the studs are housed into the plates. Locating the studs in an existing wall is essential for the fixing of a wall plate for a pergola.

TENON SAW A short, rigid-bladed saw with a steel stiffener to keep the blade dead straight. Used for critical work when cutting joints.

WALL ANCHORS Proprietary fittings that fit into pre-drilled holes in masonry and by expansion, secure varying items. A vast range of bolts, inserts and studs is available to the home handyman. Ramset and Dynabolt are two popular trade names of these all-purpose designed fixings.